HOW TO HACK AN ELECTION

DAVID & JACK CAHN

A SUTTON PLACE PRESS BOOK

HOW TO HACK AN ELECTION

ISBN: 978-1-7326848-0-5
IBSN (eBook): 978-1-7326848-1-2

Cover Design by David Provolo
Interior Design by David Provolo
Author Photograph by Isabella Cuan

Sutton Place Press
708 Third Ave, 6th Floor
New York, NY 10017

Printed in the United States of America
For more information visit: www.cahn.io

To our professors, teachers, and mentors

CONTENTS

INTRODUCTION

On a cold Sunday morning, soon after Donald Trump was elected President of the United States, I ordered my own voting machine on eBay from a government supply dealer for $150.

The voting machine arrived folded up into a neat little suitcase, with the words "AVC Edge" stamped on the front. AVC is brand of voting machine manufactured by Sequoia, one of the "big four" voting machine manufacturers, a group which also includes are Diebold, ES&S, and Hart InterCivic. These four companies collectively control more than 90% of the U.S. market for voting machines.[1]

After unpacking the voting machine, I quickly set it up in the living room of my college apartment, which I shared with my twin brother, Jack. Few onlookers, peeking through our large bay windows, knew what type of experiment we were running. In our Ravenclaw and Slytherin T-Shirts and beat-up jeans, we certainly didn't look very sinister.

Armed with the knowledge contained in this book, however, we were quite capable of having the voting machine do our bidding. Indeed, within two hours I had demonstrated three different attacks on the voting machine and posted it to YouTube. That video would ultimately form the core of my senior thesis, which this book is based on.

Why was it so easy for us to hack our voting machine? Well, to start, the hard work had been done for us. Cyber security experts at Princeton, Penn, Michigan and other research institutions have been crying foul over the numerous security vulnerabilities that plague our voting machines for quite some time now. Since 2001, the year when Congress first earmarked $3.65Bn for states to replace punch card voting systems with electronic voting machines, experts have demanded increased scrutiny of the technology, claiming than even the most amateur of hackers could do real damage to our elections.

By the time my voting machine arrived in the mail, I had read dozens of academic papers on voting machine security and learned about the internal mechanics of voting machines. I had learned about the dozens of attacks developed by academic researchers, who had often received privileged access to voting machines as part of state-mandated security audits. Armed with this knowledge, I could quickly identify key voting machine vulnerabilities and the best attacks to take advantage of them. By relying on the precedent set for me by brilliant cyber security researchers, I had developed a roadmap for hacking *any* voting machine.

Our goal in this book is to share that roadmap with you. To give you — a concerned American citizen — the context to understand why our elections are under siege, and what we can do about it. This is a book about real substance — the how and why of hacking elections. So that when the next election comes around, you understand what we're up against and what is at stake. Perhaps once you understand the issues at play here, you'll add your voice to the growing chorus of citizens demanding reform. Two-third of Americans do not have faith in the integrity of U.S. elections. They are our audience. They deserve — and indeed are demanding — an explanation for what has happened. That is the purpose of this book.

Why does this matter? Voting is a critical part of our democracy because it conveys legitimacy on our government. As evidence that our voting machines are not secure mounts, there is increasing pressure to "fix" our election system.[2] Evidence that Russia meddled in the 2016 election has transformed this problem from an academic debate into a national imperative. The Russian attacks were pervasive: hackers targeted election systems in 21 states,[3] sent 122 "spear phishing" emails to election officials, with the goal of infecting their computers with malware,[4] stole tens of thousands of emails from the DNC, penetrated the Federal Election Commission, and hacked into VR Systems, a voting machine software vendor, among other exploits.[5] A joint statement from 17 intelligence agencies confirmed that Russia meddled in the 2016 election in order to help elect President Trump.[6]

This book is not about the 2016 election, although we will discuss that topic in depth in our last chapter. Most of this book will focus on explaining in layman's terms what it means for our elections to be vulnerable, and what we can do about it. First, we'll explore specific voting machine attacks and countermeasures to those attacks. Next, we'll discuss a host of recommendations to improve voting machine security and make our own concrete recommendations. Finally, we'll describe the threat actors at play — Russia, Iran, North Korea, China and black hat hackers — and their capabilities. We'll conclude by discussing Russia's campaign to meddle in the 2016 election, and what it will mean for 2018 and 2020. Taken together, this book will paint a comprehensive picture of the state of election security today.

What to Expect

The cast of characters in this book is diverse and colorful. There are dozens of computer science professors across the country who have formed a league of sorts, dedicated to studying election security. Their approaches range from the creative to the absurd — and over the past 10 years, they have had many laughs at the expense of voting machine manufacturers. One professor got a voting machine from a disgruntled employee in a back alley. Another found the source code for a voting machine posted to an Internet blog. Yet others received voting machines as part of state mandated security audits. Together, they pioneered innovative technologies, built new friendships, and had a lasting impact on the field of computer science. They are our heroes — you'll get to know them quite well.

The villains in this story are the voting machine manufacturers: they are Diebold, Sequoia, ES&S, and Hart InterCivic. This book will reveal how these companies took billions of dollars in taxpayer money and delivered inferior voting machines. These companies failed to follow even the most basic security and software engineering protocols — protocols that even a college student is expected to follow. One company literally hardcoded the password to all their devices (the encryption key) in their code base. Once attackers stole the encryption key, they could access any and all voting machine data that was "protected." Other manufacturers incorrectly implemented basic security protocols, leaving their code vulnerable to a well-known class of attacks, called "stack-overflow" attacks. In the words of one team of researchers, "The companies that produce voting machines have poured gasoline onto the smoldering embers of concern." Their "systems have vulnerabilities that do show incompetence...And were built by programmers who simply don't know how to design a secure system."[7]

Bringing this story to life will be regular people who, through the course of their day-to-day lives, have uncovered flaws in the election process, and have shared their experiences with us. For example, Steve, an election judge, will explain how he identified gaping holes in the physical security of our elections, long before we came looking for him. He will explain how voting machines are often left completely unsecured the night before elections, giving hackers plenty of time and opportunity to cause real damage. In other cases, election volunteers, who are not vetted and are rarely subject to background checks, take voting machines home with them. Steve believes that the threat of volunteers attempting to manipulate election outcomes is material.[8]

Where possible, we include pictures and demonstrations to explain voting machine attacks, so that you can understand them without a dictionary and a search engine. We favor simplicity over complexity — without sacrificing comprehensive detail. After all, our goal in this book is to make you aware of the problem, so that you can demand that your politicians address it and fix it. Instead of hiding behind jargon, we'll blast the issues out into the open. Every American has a responsibility to ensure we protect our democracy. To fulfill that obligation, we need to understand how and why our elections are vulnerable to attack.

Along the way, you'll get to hear about some far-fetched schemes and tales. Maybe you'll even walk away with an enhanced understanding of cyber security or be inspired to get involved in the election security movement. You'll learn about how hackers turned Palm Pilot devices into universal access devices, how an elite team of Israeli hackers tricked RFID chips into letting them vote repeatedly, how one professor sued the State of New Jersey to force them to stop using insecure voting machines, how one election official identified symptoms of an attack and drew attention to a nationwide problem, and how we petitioned our local government to audit our county's election data (and you can too).

The best part about cyber security is that there are no rules. Hacking is all about learning on the fly. It requires exceptional creativity — on behalf of both the defenders and the attackers, who are constantly upping the ante, challenging each other to be better and do better. The best way to learn about defense is to start thinking like an attacker. We hope you find this book not only to be educational and interesting, but also enjoyable and fun!

CHAPTER 1

A BRIEF HISTORY OF ELECTION SECURITY

"We've gone from dead folks voting to live folks not being able to vote" — Miami-Dade Commissioner Barbara Carey-Shuler.

Our story begins in 2000, when *Bush v. Gore* revealed that minor ambiguities in America's voting system could play a major role in determining America's president. At the time, Harvard Law Professor Laurence Tribe wrote that the election was a "cliffhanger of an election, essentially a national farce."[9] Congress immediately resolved to modernize the U.S. election system and prevent flaws in the execution of elections from ever again creating such an ambiguous outcome. In 2002, the Help America Vote Act set out to achieve this goal, allocating $3.65Bn for states to replace punch card voting systems with electronic voting machines.

Congress' solution — to throw money at the problem — led to rapid adoption of Direct-Recording Electronic (DRE) voting machines (basically, computerized voting machines — the ones you used the last time you voted). By 2006, 92% of counties were using some type of electronic voting system.[10] Almost immediately, security experts warned that the machines might not be secure.[11] By and large, their concerns were ignored.

As soon as they were deployed, electronic voting machines began to cause problems. In 2002, a scandal erupted yet again in Florida when hundreds of precincts in Miami-Dade reported 0 voters. "We've gone from dead folks voting to live folks not being able to vote," Miami-Dade Commissioner Barbara Carey-Shuler, told the *Miami Herald.*

Over the next 15 years, our review of newspaper archives reveals reports of hundreds of voting machine failures across the country. Whether these failures were technical problems, or symptoms of voting machine attacks, is subject to further scrutiny. Certainly, several of these "failures" have the trademarks of standard voting machine attacks, which we will describe later in this book.

An illustrative sample of these reports is included below, to highlight just how pervasive voting machine failures have been since 2002. For those who say voting machine attacks are theoretical, and could never happen in real elections, we beg to differ.

iVotronic Machines Disenfranchise Florida Voters (2002): Chaos ensued on September 12, 2002 when election observers noticed that 0 votes had been cast in more than 100 precincts across Miami-Dade, resulting in uncertainty about who won the Democratic gubernatorial primary, Janet Reno or Bill McBride. Additionally, there were large discrepancies between the two "redundant" methods for counting votes (89 total votes according to one count, 610 according to the second, for instance). Election officials argued that the mistakes were due to 1) Poor training of precinct employees, 2) Loss of master external memory devices, resulting in delays, and 3) Modem failure, which prevented tabulation on the back-end software system. ES&S, the manufacturer, claimed that no votes had been lost. From the terminology used by election officials, they appear to have had little understanding of how the iVotronic brand electronic voting machines actually work.[12]

Memory Card Issues in Sequoia AVC Edge (2002-2004): In a series of tight elections, several counties reported problems retrieving vote counts from PCMCIA memory cards, which are used to tally the final vote count after an election. These counties included: Palm Beach County, Florida, Hillsborough County, Florida, and Bernalillo County, New Mexico, where only 36,000 votes were reported, even though 48,000 people cast ballots in the race. In June 2004, Morris County, NJ found that its PCMCIA cards had been blanked out; the chips contained a long string of zeroes, but no votes.[13]

California's Diebold Machines Ran Uncertified Software (2003): A 2003 security audit of Diebold voting machines in California discovered that Diebold machines used in 17 California counties were using uncertified software. At the time, California Secretary of State Kevin Kelley told *Wired*, "The core of our American democracy is the right to vote. Implicit in that right is the notion that that vote be private, that vote be secure, and that vote be counted as it was intended when it was cast by the voter. And I think what we're encountering is a pivotal moment in our democracy where all of that is being called into question." This incident led to the decertification of Diebold machines. However, the machines were ultimately recertified.[14]

Orange County Voters Receive the Wrong Ballots (2004): In 21 precincts across Orange County, more ballots were cast than voters. 7,000 voters appear to have cast ballots in the wrong precinct, due to a software failure in Hart InterCivic electronic voting machines. InterCivic was not able to resolve these problems, citing voter privacy issues. The company claimed that the problem was due to a failure in the machine's access control mechanism. Poll workers appear to have typed in the wrong PIN numbers when activating the voting machines. The *Los Angeles Times* reported that poll workers told some voters to write-in their candidates.[15] In other precincts, voting machines simply broke down and were replaced by paper ballots.

Candidate Preferences Ignored by iVotronic Voting Machines in Texas (2002 and 2016): In Texas, 2002 general election voters reported that when they attempted to vote for a Democratic nominee, the machine indicated that they had selected a Republican. This time,

ES&S blamed the failure on screen misalignment. "We don't know if we lost 10 votes, 100 votes, 1,000 or 10,000," Susan Hays, Chairwoman of the Dallas County Democratic Party, told the *Dallas Morning News*. Republicans claimed the error was much less pervasive.[16] Remarkably, the same issue arose during the 2016 election, according to *Dallas Morning News.*[17]

Irregularities in New Jersey Presidential Primary (2008): Irregularities were reported for 38 voting machines across 8 counties during New Jersey's 2008 primary election. In these elections, more Democratic votes were cast than Democratic voters. The opposite was the case for Republicans. An audit of the voting machines indicated that the error was caused by a frequent operator error, where the wrong ballot was activated for certain voters. This led voters to accidentally cast their ballots in the wrong primary. In the same election, a hardware bug led voter disenfranchisement. Due to improper activation of their voting machines, certain voters were incorrectly notified that their votes had been cast. In fact, their votes were never recorded.[18]

WINVote Issues Lead to Anomalies in Virginia Elections (2014): In 2014, WINVote voting machines came under attack when 57 precincts reported that their voting machines had broken due to power issues, wireless communication issues or "embedded" errors. Jeremy Epstein has called WINVote machines the worst voting machines in America and said that if they were not hacked, "it was only because no one tried." A WINVote bug that apparently subtracted one vote for every one hundred votes cast was enough to alter the outcome of at least one close race for a seat on the Fairfax School Board, *Wired* reported.[19]

59% of Precincts Report Discrepancies in Detroit (2016): In Detroit, one in three voting machines incorrectly tallied their votes in 2016, according to *Time.*[20] The inaccurate results were blamed on optical scan machines and poll volunteers.[21] Michigan law states that instead of resolving discrepancies, ballots from precincts with discrepancies are not counted during a recount. This means that hacking a voting machine can directly result in voter disenfranchisement. In January 2017, Detroit announced it would spend close to $100MM on new electronic voting machines from ES&S, Dominion (formerly Sequoia) and InterCivic.

Corrupted Memory Cards in North Carolina ES&S M100 Optical Scanners (2016): In the hotly disputed 2016 gubernatorial race between Pat McCroy and Roy Cooper, McCroy's campaign claimed that 90,000 ballots had been counted incorrectly, according to the *Raleigh News & Observer.* The memory cards, responsible for storing the final vote count, appeared to have "experienced a critical error" during vote processing, due to corrupted memory chips.[22]

AVC Machines in Pennsylvania Switch Some Votes, Ignore Others (2016): CBS News reported multiple issues with Pennsylvania voting machines, including machines that were switching ballots in Clinton Township and machines that failed to record votes in Allegheny County. In Philadelphia, certain polling locations reported long lines due to machine malfunctions. Election officials said calibration issues were to blame for the failures.[23] (Your authors both personally waited three hours to vote in Pennsylvania in 2016, due to voting machine failures. Former Pennsylvania Governor Ed Rendell handed out donuts to encourage predominantly liberal Philadelphia voters not to leave the voting line).

Memory Chip Failures Disable Utah Voting Machines (2016): Voting machines in southern Utah appeared to have failed due to glitches in the memory chips, resulting in the use of paper ballots, according the *Chicago Tribute.*[24] In Washington County, Utah, a full 75% of voting machines failed, according to County Clerk Kim Hafen, in an interview with *Vice.*[25] Officials did not disclose the technical details of the issue, stating only that the problem was caused "improperly coded memory cards."

U.S. Election Commission Hacked (2016): The Election

Assistance Commission (EAC), the body responsible for certifying and maintaining electronic voting machines for use in U.S. elections, confirmed in December 2016 that it was hacked earlier in the year. According to Recorded Future, a cyber security firm, the breach exposed the usernames and passwords for more than 100 EAC employees. The attackers were caught attempting to sell those credentials on the black market. The attackers — who have been code-named Rasputin — apparently leveraged a simple SQL injection to steal the credentials. SQL injections are one the most naïve attacks on the Internet and involve typing code directly into a search bar, causing a supposedly hidden database to print or delete its contents.

In conclusion, repeated failures of electronic voting machines over the past 16 years and especially in 2016 provide tangible evidence that the security vulnerabilities identified by researchers may pose a real threat to election security. Many of the issues noted above directly correspond to certain attacks we will discuss, such as a screen calibration attack (Texas), a targeted denial of service attack (Detroit), and an access control attack (New Jersey). If nothing else, the evidence we have accumulated from reviewing newspaper archives demonstrates that voting machine failures are not an exception, but rather the norm.

Our final example, the 2016 hack of the U.S. Election Assistance Commission (EAC) should raise a huge red flag.[26] If the election officials themselves are vulnerable to such basic attacks, then we should not be surprised that two-thirds of Americans do not believe their votes will be counted as cast.[27] There is clearly a significant amount of work that needs to be done to restore faith in our elections.[28] In the coming chapters, we will contribute to this important conversation.

CHAPTER 2
EASY AS A BUTTON

"If we do not change the process of designing our voting systems, we will have no confidence that our election results will reflect the will of the electorate. We owe it to ourselves and to our future to have robust, well-designed election systems to preserve the bedrock of our democracy."
— Kohno et al.

The most straightforward way to hack an election is to disable all voting machines. Another simple approach would be to cast many duplicate votes. This chapter is focused on how a hacker might manipulate flawed access control rules or broken user interfaces to achieve these goals. We consider three equally potent attack scenarios.

In scenario one, the hypothetical attacker casts 10,000 votes instead of a single vote. Thus, a group of attackers could change the outcome of a local or congressional election. With enough manpower, a state-actor could change the outcome of a swing state, and thus influence a presidential election. China and Russia are both known to have extensive cyber warfare capabilities, including entire units of their armies dedicated to cyber warfare. For practical purposes then, manpower is no challenge for state-actors, nor is funding. The only roadblock to a successful attack would be rigorous defenses. As we will show, voting machine defenses are extremely weak.

In scenario two, the attacker changes ballots while an election is in progress. For example, an attacker alters the screen calibration system, so that the button for "Hillary Clinton" leads the computer to add "+1" to its Donald Trump vote counter. As you saw in a Chapter 1, screen calibration issues have consistently been one of the most commonly reported problems during actual elections. A simple Google search reveals dozens of reported examples in 2016 alone. Alternatively, an election volunteer can load the wrong ballot for a voter — for example, give Democratic voters the Republican primary ballot, or simply fail to enable voting in the first place. These decisions would lead to widespread voter disenfranchisement. As we shall see, this is exactly what happened in New Jersey, resulting in a noisy legal battle that pitted bureaucrats against cyber security researchers, and ultimately shed light on just how broken our election system has become. When voting machine manufacturers refused to cooperate in the New Jersey investigation, citing trade secrets, they revealed a callous disinterest in the security of their products and the integrity of the elections they help operate.

In scenario three, the attacker focuses on a "denial of service" attack. Denial of service involves disabling voting machines, to create long lines and decrease voter turnout, or tampering with voting machines in detectable ways, so that their votes get disqualified by state election authorities. As we learned in 2016, when security researchers supported a recount in Wisconsin (successful) and Michigan (unsuccessful), state election laws err on the side of disqualifying ballots, rather than redoing an election. We will argue that denial of service is the most actionable, and therefore the most worrisome threat to election security today. Because voter preferences are well known, the most effective way to influence election outcomes is to simply stop voters who disagree with you from voting at all. In Utah, where voting machines broke, or in Philadelphia, where we personally experienced 3+ hour lines, due to broken voting machines, we see symptoms of denial of service.

In this chapter and subsequent chapters, we'll focus on painting a detailed picture of each of these attacks, and we will build threat models that help us contextualize them. We'll bring to life the technical research and political drama that has surrounded these topics for more than a decade. And we'll help introduce you to the research teams that invested years of sweat and hard work into identifying voting machine vulnerabilities, so that we can stay one step ahead of the hackers who hope to subvert our elections.

Attack #1: Access Control Attack on iVotronic Voting Machines
Research Team: Aviv et al. (2008)[29] and McDaniel et al. (2007)[30]
Target Machine: iVotronic by ES&S

iVotronic voting machines are electronic voting machines currently deployed in 20 states. They are manufactured by ES&S, the world's largest e-voting machine vendor. ES&S machines — including the iVotronic model (97,000 units) and the M100 Optical Scan (30,000 units) — are used in counties that represent more than 67 million voters.

Despite their prevalence, ES&S voting machines are vulnerable to a series of hardware attacks that manipulate the machines' Personalized Electronic Ballots (PEB), which are special hardware devices responsible for access control. Access control generally refers to the idea that registered voters, and only registered voters, should be given a single vote. The PEB is a palm pilot-like device used by poll workers to reset the voting machine for each voter and display the correct ballot. The PEB is also used after the election to close the election and send all ballots to a central voting system, called Unity. Unity tallies up the ballots and is responsible for reporting county-wide winners. PEBs are also memory storage devices. Every time a person votes, the iVotronic machine writes the vote down to a block of memory in the PEB. The PEB will eventually transmit this information to Unity.

Although PEBs are vulnerable to a wide-variety of attacks, ES&S did implement several security measures to protect them. There are four primary defense mechanisms for PEBs:

1. Administrative functions are password protected
2. For mission critical functionality (i.e. firmware) and sensitive memory (i.e. ballot results), an Election Qualification Code (32-bit number) is required for modifications.
3. All voting inputs to the iVotronic machines are recorded via a Real-Time Audit Log Printer (RTAL). When suspected improprieties are uncovered, the Voter Verifiable Paper Audit Trail (VVPAT) can be used to investigate what happened and confirm that votes were tallied correctly.
4. ES&S issues two types of PEB devices, each with different levels of access: the blue Voter PEB and red Master (or "Supervisor") PEB.

On face value, these defense mechanisms are appropriate. Their flaw, however, is improper threat modeling. ES&S's theoretical defenses don't consider the mindset and capabilities of their potential foes. This makes the machines vulnerable to basic attacks that do not require any software engineering expertise.

The most naïve attack on a PEB-based iVotronic machine is to spoof a PEB-device. This is a relatively simple process because there are only two unique features that specify a PEB: the ability to communicate via IR protocol (infrared electromagnetic spectrum) and a magnetic detection system. Though PEB sounds like a very specific device, it functions as a general class of handheld communication devices. As a result, it is not difficult to emulate a PEB by programming a Palm Pilot (which communicates via IR protocol) to simulate PEB-like communication and use a small magnet so that the iVotronic detects the Palm Pilot as being a PEB. In **Figure 1**, a researcher demonstrates a PEB-emulator.

With a PEB-emulator, a hacker can have a material effect on the outcome of an election. Here is a sampling of potential attacks:

Modify Screen Configurations: Using a PEB, an attacker can re-calibrate the iVotronic screen and deny voters access to certain parts

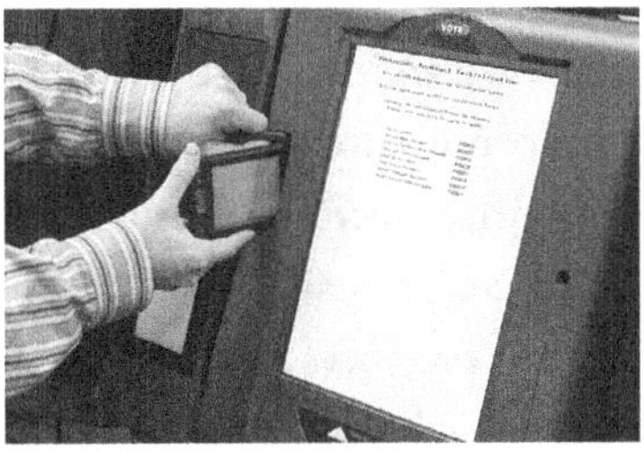

Figure 1: *A researcher uses a Palm PEB-simulator to update configurations on an iVotronic machine.*

Figure 2: *A researcher demonstrates how an iVotronic VVPAT audit log could be disabled by unplugging it.*

of the screen. While this attack is likely to be spotted rather quickly, it has a few benefits. First, it takes a few seconds and will not be conspicuous. Second, it could be used to cause delays and confusion at critical voting locations, especially if the poll workers are not tech savvy. An attacker might target heavily Democratic or Republican districts in a swing state to reduce voter turnout, for example. Former CBS News Anchor Dan Rather has claimed that there is "conclusive evidence" that there have been irregularities consistent with such an attack during actual elections.[31]

Erase Vote Counts: Sensitive PEB-features are password protected. However, passwords can easily be bypassed with a special type of PEB, called a Quality Assurance (QA) PEB. QA PEBs are just as easily spoofed as regular PEBs. In fact, the difference between a regular PEB and a Quality Assurance PEB is a single character in the communication protocol. Passwords are the only defense for many sensitive PEB functions, including "clear-and-test" which will erase all existing votes on the targeted voting machine. Thus, an attacker with a spoofed PEB (a re-programmed palm pilot) can delete all the votes for a voting machine at the end of the election day, thus disenfranchising targeted voters.

Alter Vote Tallies: So far, our threat model has assumed that the attacker is a registered voter. If we modify that model to assume that a poll worker could be acting maliciously (which is possible given that many poll workers are volunteers and do not have to go through a vigorous vetting process), then the attack surface increases substantially. Specifically, anyone with access to a precinct's main PEB (i.e. the supervisor PEB) could potentially alter their precinct's results.

First, the attacker would need to get through an additional layer of defense: the Election Qualification Code (EQC). This code is used to encode a cryptographic key, which is in turn used to encrypt the election results. The weak link in this system is the EQC, which is not password protected. Anyone with access to the Supervisor PEB, a laptop computer and a small magnet can extract the EQC. Using the EQC, the attacker can write arbitrary memory to the PEB device — including altering the vote tallies. It would not be suspicious for a poll worker, who uses the PEB in his/her daily activities, to sit down with a laptop and the PEB at some point and write the short amount of code necessary to execute this attack. Thus, any insider, with a laptop, a magnet, and a few uninterrupted minutes, can re-write all the saved votes for their election precinct.

Stop an Audit: What about an audit? You might imagine that any of these attacks might lead authorities to conduct a recount (this happened in Wisconsin in 2016, for example). In this case, authorities would attempt to verify the election results using paper ballots, which are printed by the voting machine after each ballot is cast. If the attackers' goal is simply to cause disruption (and thereby decrease voter turnout), then the audit is irrelevant. However, disrupting an audit would be very straightforward. As illustrated in **Figure 2** an attacker would simply need to unplug the real-time audit log (RTAL) printer, which is not protected. In conjunction with PEB spoofing, this would be at minimum disruptive, and at worst, could lead to the invalidation of all votes cast at a target precinct. However, attackers can do better than just unplugging the printer. With additional hardware (a smartphone or iPad), a sophisticated attacker could use the open printer connection to print arbitrary paper ballots, which would be indistinguishable from valid votes. Finally, an attacker could simply destroy the audit log by pouring a chemical over the printer. This would either destroy the printer or its ballots.

Conclusion: The most dangerous element of the physical PEB attacks we've discussed thus far are not the components, but the sum of the parts. Any individual attack — for example, spoofing a PEB and changing the configuration settings, or clearing the votes on an individual machine — are limited in their potential impact. However, when combined, these individual vulnerabilities create major security threats that we are likely incapable of handling. For example, a sophisticated attacker could (1) unplug the real-time-audit-log (RTAL) printer and connect his/her printer, which would print arbitrary paper ballots (VVPATs), and then a second attacker (installed as a poll worker) could add these votes to the PEB after the election is complete using the Supervisor PEB. This confluence of attacks would yield consistent results in the digital log (sent to Unity) and in the paper audit trail.

In the next chapter, we'll address other weaknesses in ES&S's PEB-based voting system, including malware and firmware vulnerabilities. As we shall see, these hardware attacks only scratch the surface of potential PEB vulnerabilities. Software and firmware attacks could do far more than just corrupt a single precinct or voting machine; they could be used to take over the entire Unity system, which aggregates results across counties, and create persistent malware to modify countywide results in more than one election.

Attack #2: Defeating Diebold's Smart Cards

Research Team: Kohno et al. (2004) [32]

Target Machine: AccuVote-TS by Diebold

In 2004, shortly after electronic voting machines were adopted nationally, Tadayoshi Kohno, Dan Wallach and a group of security researchers discovered the source code for the Diebold AccuVote-TS direct electronic voting machine on the Internet. Allegedly, the code was posted to an online repository by accident. What followed was a groundbreaking paper on electronic voting machine security that catapulted its authors to positions of prominence is the security research community. Wallach, especially, became a figurehead for a new movement of security researchers who would spend the next decade hacking voting machines and proving that direct electronic voting machines were dangerously easy to manipulate.

"We show that voters, without any insider privileges, can cast unlimited votes without being detected by any mechanisms within the voting terminal software," the researchers explain. "Our analysis shows that this voting system is far below even the most minimal security standards applicable in other contexts." The source code was so damning that Wallach and team concluded that privately written code — by any vendor — would be "unreliable." "If we do not change the process of designing our voting systems, we will have no confidence that our election results will reflect the will of the electorate," the team concluded. "We owe it to ourselves and to our future to have robust, well-designed election systems to preserve the bedrock of our democracy."

Wallach's team initially focused on access control vulnerabilities created by the use of credit-card sized smart cards, called Voter Access Cards which are inserted into the Diebold machine to enable each voter to cast his/her ballot. When a voter enters the polling location, s/he receives a Voter Access Card and inserts it into the voting machine. When the machine registers the presence of the smart card, a ballot is loaded onto the screen and the voter selects his/her favored candidates. At the end of voting, the smart card is returned to a poll worker, who re-programs the Voter Access Card so that it can be re-used, using a device pictured in **Figure 3**.

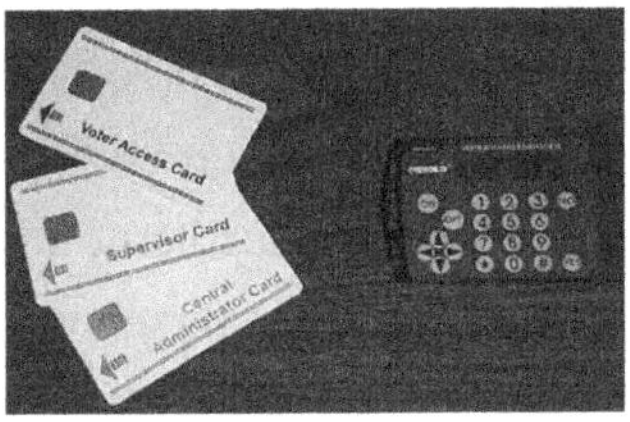

Figure 3: *Diebold smart cards along with a smart card programming device*

	Voter (with forged smartcard)	Poll Worker (with access to storage media)	Poll Worker (with access to network traffic)	Internet Provider (with access to network traffic)	OS Developer	Voting Device Developer
Vote multiple times using forged smartcard	•	•	•			
Access administrative functions or close polling station	•	•			•	•
Modify system configuration		•			•	•
Modify ballot definition (e.g., party affiliation)		•	•	•	•	•
Cause votes to be miscounted by tampering with configuration		•	•	•	•	•
Impersonate legitimate voting machine to tallying authority		•	•	•	•	•
Create, delete, and modify votes		•	•	•	•	•
Link voters with their votes		•	•	•	•	•
Tamper with audit logs		•			•	•
Delay the start of an election		•	•	•	•	•
Insert backdoors into code					•	•

Figure 4: *A quick summary of the attacks possible against a Diebold voting machine, based on type of attacker*

In addition to their role as access control devices, Voter Access Cards store basic data, such as the political party of the voter, which is used to determine which ballot needs to be presented to the voter during a primary election. Voter Access Cards are not the only type of smart card used by Diebold. The system comes with Supervisor Cards and Central Administrator Cards as well, which are used by poll workers to end the election when voting is completed and begin the process of tallying the votes. Ending the election triggers the Diebold machine to write the election results to removable flash memory and/or transmit the results to Diebold's back-end server, which is located at county election headquarters.

Diebold's smart cards lack cryptographic protection. Cryptography is usually used in computer science to guarantee data integrity. Hashing, encryption, and digital signatures are three common cryptographic tools. Cryptography converts meaningful information into a "cipher." Only parties with the correct key (trusted parties) can decode the cipher, thus ensuring that data is secure. In the case of digital signatures, the correctly encoded cipher is used to guarantee that the correct certification authority produced the software and you are not being hacked. In the case of smart cards (like credit cards), cryptography is needed to make sure the smart card was issued by a trusted authority, not by an attacker in his garage.

Diebold's lack of encryption is surprising since one of the key benefits of smart cards is that they can use physical (and therefore hard to steal) keys to verify card integrity. Diebold chooses not to do this. A quick Google search indicates that buying programmable smart cards on the Internet is affordable. I can buy a set of 5 programmable smart cards for $12.90 on Amazon or a set of 8 for $11.02 on eBay.[33] Programming "homebrew" smart cards to circumvent the access control mechanisms on a Diebold voting machine is easy and affordable.

To create a homebrew smart card, the hacker would need to learn how to code in the specific protocol used by the Diebold voting machines. We believe that this is the only real challenge posed to an amateur hacker. Wallach et al. explains a few mechanisms for learning this protocol. For example, one scenario involves a duo "probing" for the protocol, in advance of their attack. The first hacker (posing as a voter) could line up to vote. At the end of voting, he will return his smart card to the poll worker, which is normal operating procedure. However, the card he returns will not be the card he received. Instead, he will pull a "probing smart card" out of his pocket, pre-programmed for this attack. When the poll worker goes to refresh the smart card for use by a subsequent voter, the smart card will record this protocol. The next attacker (again posing as a voter) will be next on line, positioned to receive what he knows to be a "probing smart card." The unwitting poll worker will hand him the card, thinking it to be a regular smart card. Instead, he will be handing over key protocol information to the attackers, who can use it to launch a future double-voting attack.

Stealing the protocol is a one-time cost; once this process is complete, all 33,000 Diebold voting machines will be exposed to homebrew attacks until they are decommissioned. This is critical to our threat model: since the primary challenge is stealing the Diebold protocol, and all Diebold machines use the same protocol, a rational attacker will be willing to invest the time and effort needed for the attack above, to gain access to 33,000 voting machines — indefinitely.

Thinking like an attacker, we would probably execute this attack at a remote polling location during an off-cycle election (i.e. city council) race. This ensures that the attackers receive their "probing smart card" back, and not someone else and will also guarantee minimal scrutiny. For all we know, attackers (i.e. Russia) have already stolen the Diebold protocol and homebrew smart cards are in use today. Such an attack would be difficult to detect.

A slightly more advanced way of learning the communication protocol that is used by the smart cards involves a "wiretap" mechanism. Usually, tapping into communication between two devices (i.e. smart phones) is not possible because the messages are encrypted. Fortunately, voting machine communication on the Diebold machine is unencrypted. Therefore, to effectively wiretap the voting machine, the attacker need only set himself up as a "man in the middle," intercepting communication between his own (legitimate) smart card and the voting machine, observing all of the communication while it happens. The attacker need only bring a smart phone to serve as the "man in the middle" and record the communication.

Once the attacker steals the communication protocol used by the voting machine, he will need to design "homebrew" smart cards that speak the same language as the legitimate smart cards. To initiate this communication, the attacker's card will need to transmit the correct passwords to the voting machine. This is how the Diebold machine is designed to distinguish legitimate Voter Access Cards from forged ones. Unfortunately, this is an extremely weak security protocol: the legitimate smart cards store the correct passwords in plain text. Thus, the passwords can easily be copied onto forged smart cards. (It is quite absurd that Diebold uses plain text passwords rather than digital signatures to secure voting machine communication, since the entire purpose of smart cards is that they can effectively store private keys in their hardware).

In any case, Wallach and his team discovered that Diebold voting machines use three hard coded passwords to check if they are receiving a legitimate smart card. These passwords are labeled m_ElectionKey, m_VCenter and m_DLVersion. To complete this attack, the attacker would need to steal these values. M_ElectionKey and m_DLVersion are election specific values, so they only need to be stolen once. If m_VCenter is unique for each polling location, then attackers will need to steal it from each voting location that they plan to attack. Because stealing the passwords from smart cards requires probing them for the data, this attack is viable for insiders, such as poll workers. Note that m_VCenter may be formulaic, which would greatly simplify this attack, because you could change the encoding based on the formula for each precinct you decide to attack.

The availability of "homebrew" smart cards enables attackers to cast multiple votes. The attacker needs to bring a stack of smart cards to the voting booth (or a single smart card programmed to never deactivate). He or she will then insert the smart cards sequentially and cast as many ballots as possible (each vote takes approximately 5 seconds) before leaving to avoid raising suspicions. Each individual attacker could potentially cast a hundred votes (five minutes is a reasonable amount of time to spend in a voting booth) and a group of attackers could magnify the impact of this attack to cast thousands of illegitimate votes.

Although this attack requires significant manpower to be effective, it is decidedly low tech and therefore easy to implement. A single attacker could manufacture and distribute the smart cards and then enlist collaborators to use the cards. This attack is detectable if election officials notice that there is a difference between the number of votes cast and the number of voters who voted on Election Day. However, realizing this would require an audit of the voter sign-in book and, separately, Appel et al. notes that these audits are rare. If the attack does get detected, then it will function as a denial of service attack because election officials will have no way of determining which ballots are legitimate and will therefore need to vacate all votes cast.

Forging an Administrator Card or Supervisor Card is no harder than forging a Voter Access Card. In fact, an analysis of Diebold's code indicates that the only difference between these cards is a single byte value. The m_CardType value set is set to 0x01 in the Voter Access Card, 0x02 in the Supervisor Card and 0x04 in the Administrator Card. A programmer can easily change these bytes when programming his/her forged cards. Unlike a Voter Access Card, Supervisor and Administrator Cards require their users to input a PIN number, which is hard coded into the card. This is no problem for our attackers. Since the attackers are themselves programming the card, they can hard code a PIN number into these cards and then write it down on the surface of the card for future use. Even if this were not the case, stealing or guessing the PIN for a legitimate Supervisor/Administrator Card is trivial. The PIN is transmitted in plaintext and is four digits long, which means a brute force solution (which simply tries every permutation of a 4-digit number) will find the PIN in a matter of seconds.

Use of an Administrator Card enables simple denial of service attacks because it allows the attacker to end the election before the end of the day. Once an election is "over," the only way to make the Diebold DRE accept votes again is to restart the election (and go through all the initial setup steps). This has the potential to 1) destroy existing votes and 2) deter legitimate voters by creating long lines. Denial of service attacks can affect elections, especially when certain precincts have known preferences (historical election results are available online on a precinct by precinct basis). Many states do not require voters to present photo ID before voting (i.e. California), which decreases the probably that attackers will be caught.

Attack #3: Relay Attacks on RFID-Enabled Smart Cards

Research Team: Oren et al. (2009) [34]

Target Machine: Any RFID-enabled voting machine (i.e. Israel, California)

In contrast to Diebold's insecure smart cards, some companies have proposed RFID-enabled smart cards as a potentially more secure access control mechanism. RFID-enabled systems ensure proximity between the voter and the voting machine. Though they are not used extensively, RFID-enabled smart cards are gaining popularity. Alameda County — which is the 7th largest county in the United States and includes much of the San Francisco Bay Area — is the largest U.S. County to have adopted RFID-enabled smart cards. RFID systems are on the rise globally; RFID voting has been used in India, Nigeria, Austria and Israel, for example.[35]

Proponents of RFID-enabled voting machines argue that they are more consistent and more reliable than standard voting machines. [36] Nonetheless, there are practical reasons that this system is unlikely to become widespread in the United States. RFID-enabled voting is not only costly to implement, but it also imposes potential privacy issues. We won't focus on these issues here. Instead, we'll explain how an attacker could use what is known as a "relay attack" to defeat the RFID system and manipulate election results.

An RFID system consists of three parts: a card reader, a card, and electronic tags associated with each card. The reader is activated by magnetic coupling. When the card is within 5cm of the reader, the reader will get activated. Think of the cards used to swipe into hotel rooms, or into your office buildings. These cards often use RFID technology.

A relay attack is a special type of man in the middle attack. Remember that in computer science, a man in the middle attack involves a malicious party standing in between two legitimate parties (i.e. the card and the card reader) and intercepting their messages. The attacker can either be passive or active. A passive attacker merely reads the messages before forwarding them to the intended party. An active attacker alters the messages to manipulate the communication mechanism. As its name indicates, a relay attack involves stealing messages from a legitimate tag, and then sending them to the card reader, so it believes you are presenting it with a legitimate RFID tag. Really, you are presenting it with the relay device. Thus, you circumvent the RFID protection.

Oren et al. detail a specific type of relay attack, which they describe as using a "leech" and a "ghost." The leech is used to convince a victim "tag" (or card) that it is a reader. Thus, it latches onto a legitimate tag. The ghost card presents this tag to the victim reader, convincing it that it is legitimate. A relay attack can be executed even if there is strong encryption because the leech is passing a *legitimate message* from the victim to the reader. Experimentally, Hancke et al. built an RFID system that extends the range of an RFID system by 50 meters. [37] Since the entire purpose of RFID chips is to ensure physical proximity, this effectively renders the system useless. The card being presented to the card reader (and thus, the poll worker) could be using an electronic tag from a voting card you don't own. Thus, you have stolen votes.

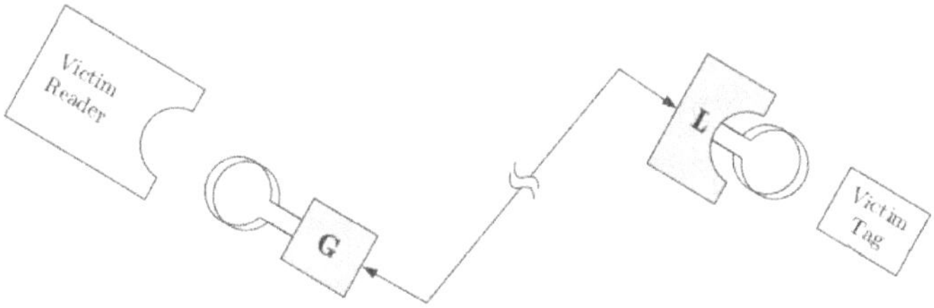

Figure 5: *An RFID Relay Attack. G is the "Ghost" and L is the "Leech." Diagram from Oren et al.*

Relay attacks have been extensively studied. Oren et al. highlights two examples of these attacks: the "Chess Grand Master Attack," and the "Mafia Fraud" attack. In the Chess Grand Master Attack, a computer that does not know how to play chess can simultaneously play against two grandmasters, such that they both believe it knows how to play chess. It does this by relaying the moves of both grandmasters to each chess board. The Mafia Fraud attack is used to defeat Fiat-Shamir authentication, a digital signature protocol.

Let's now apply the relay attack to hacking an election. Oren et al.

are focused on precisely this application. In fact, the purpose of their original academic paper was to provide evidence that RFID-based voting was not secure, and should not be adopted by the Israeli government, which was debating a proposal to do so. The first consideration in using a relay attack to hack an election is to avoid detection. This is rather trivial; Oren et al. propose that an attacker could simply hide the leech and ghost in his/her sleeve. The attacker could also put them in a pocketbook or bag.

To demonstrate how relay attacks could be used in practice, Oren et al. construct a mock election. The set-up of their hypothetical election is based on the system used in Israeli elections: voters present their cards at the voting booth and make a candidate selection, which is saved down to the smart card. Before leaving, the voters drop their smart cards into a "ballot box." To confirm that the correct voting information was saved to their smart cards, voters could swipe their card at a "verification terminal," which would display the choices saved to the smart card. They duplicated this set-up in the lab, to demonstrate a series of security vulnerabilities.

Ballot Sniffing Attack: This attack is used to steal the current vote tally before it is officially announced, by leveraging the voter verification terminal. Using the verification terminal, the Oren et al. team set up a relay between the smart cards physically stored in the ballot box and the verification terminal. The relay attack is particularly potent here because in the privacy of the voting booth, the attacker's ghost/leech is unlikely to be detected. The outcome of this attack is that the hacker could see what votes had been cast. To actively effect election outcomes, a ballot sniffing attack would be used in combination with a denial of service attack, discussed below. The combination allows the attacker to specifically destroy ballots on voting machines that favor his/her opposing candidate.

Single Dissident Attack: The threat model for this attack assumes that there are "dissidents" who want to manipulate an election result for their favored party. Using heuristics about voters (such as age, race, gender) these dissidents target certain voters to suppress their votes. To execute the attack, the dissident sets up a relay between a single vote that has already been cast and the voting/verification system. Thus, when targeted voters intend to cast their ballots, the "single dissident ballot" is displayed instead. In fact, the voter's ballot remains blank and does not get cast. There are two prerequisites for this attack. First, the attacker must be physically present (and therefore must be a volunteer at the voting location), and second, the attacker must be very certain about the voting behavior of his target audience (otherwise, he will display the wrong fake vote and raise a red flag). There is a clear countermeasure to prevent this attack. After each voter is finished, his/her smart cards should be disabled.

Ballot Stuffing: RFID-based voting machines are designed to allow re-writing ballots (the whole point of the verification system is to allow voters to correct their mistakes). This means that an attacker who sets up a relay between previously cast votes in the ballot box and the voting machine can change previous votes. Like with the single dissident attack, disabling cards once they have been used and verified would prevent modifying ballots after the fact.

Based on the evidence presented above, it does appear possible to develop a secure RFID-based voting system. The vulnerabilities present in existing RFID-based voting machines are correctable. Though they are not perfectly secure, RFID-based systems may provide enhanced security relative to the status quo.

There are also non-relay hardware attacks that may be used against RFID-based voting systems. These include:

The Zapper Attack: This attack uses a high-power electric pulse to break the RFID chips and overwhelm the cards' circuitry. An attacker might execute such an attack to disqualify votes from a precinct with

Figure 6: *A prototype for an RFID-zapper developed by German engineers.*

known political preferences. However, like with the relay attacks above, there is a known countermeasure to prevent this attack. Smart cards can be designed with circuitry that resists static electricity stocks. Unfortunately, the cost of this defense mechanism may be prohibitively expensive. The primary concern with using an EMP is that it might be conspicuous. However, re-searchers in Germany have demonstrated that this can be done with a camera's flash, as you can see in **Figure 6**.[38]

Denial of Service: An attacker can execute a denial of service attack using an antenna, which will broadcast a random SSB-modulated bit pattern at a frequency of 14.47 Hz. This frequency will confuse the RFID reader, which will be unable to distinguish between the actual RFID card and the radio frequency. Likewise, a "blocker tag" provides another vector for initiating a denial of service attack. A "blocker tag" is a specifically engineered smart card that is programmed to circumvent the reader's anti-collision protocol and therefore confuse the RFID reader. In practice, what would happen if voting machines suddenly stopped working? It is likely that poll workers would not be able to fix the machines and would need to call in experts. This would take quite a bit of time and deter targeted voters, at the very least. Countermeasures against these attacks include blocking radio signals and designing an anti-collision protocol that cannot be experimentally defeated.

Attack #4: User Interface Attacks on the Sequoia AVC Advantage

Research Team: Appel et al. (2009) [39]

Target Machine: AVC Advantage by Sequoia

In 2004, Rutgers Law School Constitutional Litigation Clinic sued the State of New Jersey, claiming that the use of direct-recording electronic voting machines (DREs) in 18 out of 21 New Jersey counties was 1) illegal because New Jersey statute guarantees that all votes be counted accurately, and 2) unconstitutional because the New Jersey Constitution says that all votes must be counted (i.e. not ignored). [40] The case was dismissed when the New Jersey Legislature passed a bill requiring that all New Jersey voting machines keep paper backups of their votes (at the time, Sequoia's AVC machines relied on firmware alone to count votes).

In 2008, after years of delay in adding paper ballots to New Jersey's elections, Judge Linda Feinberg of New Jersey's Superior Court ordered a trial to determine the constitutionality of paperless ballots. The plaintiff called on Princeton Professor Andrew Appel as an expert witness and the court mandated that Sequoia provide Dr. Appel with a voting machine for a full-fledged security audit. As a result, the security (or lack thereof) of the AVC Advantage is well documented. Appel's testimony was the first of its kind and has since prompted a similar lawsuit in Pennsylvania, which also uses AVC Advantage voting machines.

Appel takes advantage of two software engineering issues to take control of AVC voting machines. First, he leverages design flaws. Software design is the study of best practices to improve the internal quality of software (i.e. testability, maintainability). Second, he leverages testing errors — program bugs that were not identified during the development cycle as "back-doors" into AVC's firmware. Much of the code for the AVC Advantage is written in Z80 assembly language and in C. In this section, we'll focus on attacks that do not require any code. These attacks leverage design flaws in the hardware alone. We'll return to the software issues later in Chapter 3.

The AVC Advantage has two primary design flaws in its user interface which has led to voters being disenfranchised by careless poll workers. In practice, an attacker could take advantage of these design failures to systematically disenfranchise target voters.

1. Machine Activation is Ambiguous: The AVC Advantage requires poll workers to press "Activate" before a voter casts his/her ballot. However, even without an activated ballot, the machine will still appear to record the vote, and display "Vote Recorded. Thank You" to the voter: This occurs even though the ballot is not activated, and the vote was not recorded. On the flip side, it is possible to cast too many ballots. For example, the only feedback for an activated ballot is a short chirping sound when the ballot is activated. If the noise level at the polling station is high (which it often is) the sound may not be heard. In this case, a poll worker may ask the voter to vote again. These accidental mistakes that may affect election results. For example, Precinct 6 in Pennsauken, NJ results showed a 1% "under-vote" (i.e. there were 1% fewer votes cast than voter stubs counted). This issue has important constitutional implications, since every American has a constitutional right to have their vote counted.

An even more dangerous attack is possible if our threat model assumes that poll workers may be attackers. No full-fledged study has analyzed the vetting process for polling station employees, but anecdotal evidence indicates that it may not be rigorous. We assumed earlier in this book that it might be possible to distinguish the voting preferences of certain groups of voters based on external factors such as age, gender, race and religion. A poll worker who is an attacker would not find it difficult to systematically "forget" to correctly activate the machine.

This attack has three implications. First, there's no way to correct the issue (absent a re-vote). Since individual votes are blind, the State will only be able to identify the number of disenfranchised voters, not their identities. Second, since there are multiple poll workers operating the machines simultaneously, identifying the culprit would be difficult. Finally, the State would be unlikely to get a conviction, even if they did identify the culprit, because distinguishing malicious behavior from ineptitude is difficult, especially when the system is set up to make activating ballots ambiguous.

Figure 7: *Intentionally or accidentally, misuse of the operator panel can disenfranchise voters*

2. Pressing the Wrong Button is Catastrophic: Evidence from the 2008 presidential primary election indicates that software bugs in the AVC Advantage led to 1) the disenfranchisement of voters, and 2) incorrect result tallies. During this election, Union County Clerk Joanne Rajoppi noticed that the AVC machine's total votes cast value was not equal to the sum of the candidate values. Based on her findings, clerks across New Jersey looked for similar problems. All in all, 38 machines in 8 counties were shown to have the same problem. The bug had to do with the order in which the poll worker pressed the "activation" buttons. If the poll worker pressed button 6 (for Democratic Primary) and then Activate, the ballot displayed correctly. However, if any other button press preceded Activate (i.e. 6-8-Activate) due to operator error, then the machine would log this as a Democratic vote, but would activate the Republican ballot.

This is illegal according to New Jersey law for two reasons. First, registered Democrats were not allowed to vote in the Democratic primary since the Democratic ballot was not activated and second, registered Democrats had the ability to vote for Republicans (since the Republican ballot was activated). It is unclear if bug fixes have been pushed to machines in New Jersey and elsewhere. Furthermore, as noted above, this is an ideal environment for a malicious attacker who doesn't need write a single line of code. The attacker need only show up and "accidentally" press the wrong buttons to disenfranchise certain groups of voters.

Conclusion: Hardware flaws on the AVC Advantage appear to be minor, but in practice, can be magnified by a sophisticated attacker to alter the results of an election.

Attack #5: JBC Attacks on Hart InterCivic eSlate

Research Team: Proebstel, et al. (2007) [41] and Butler et al. (2009)[42]
Target Machine: eSlate by Hart InterCivic

In 2006, Yolo County election officials reached out to security researchers at nearby University of California, Davis for help training its staff (mostly retirees) in using and setting up eSlate voting machines. Thirty or so volunteers, mostly undergraduate students and graduate students working in security research, aided Yolo County on Election Day. In return, Yolo County provided the researchers with six Hart InterCivic eSlate machines — which were enabled with audio for blind voters — so that they could be studied and security vulnerabilities could be identified. Proebstel's team did not have access to source code or machine protocol and was therefore experimenting on a "black box" system. Even so, they found serious vulnerabilities which pose an immediate threat to election security.

The Hart InterCivic's eSlate is one of the dominant voting machines models used in the United States. These machines are predominantly used in Ohio, California, Colorado, Idaho and Hawaii. eSlate machines require extensive set-up before the election, including generating a master cryptographic key for use in all voting machines for a county, and pre-loading memory cards (PCMCIA cards) with ballot definitions. Hart calls the cards "Mobile Ballot Boxes," since they are ultimately transported to county headquarters and used to tally votes. Mobile ballot boxes are initialized using a software system called BOSS, which contains the candidate database. Each eSlate machine is paired with a Judge Ballot Controller (JBC), which is used by poll workers to enable eSlate machines for each voter and to load the machine with the correct ballot. The JBC is thus our central access control device, analogous to Diebold's smart cards and iVotronic's PEBs.

The eSlate can be configured as an optical scanner, which scans and records paper ballots, or as a traditional electronic voting machine with a computer-like interface. The optical scanner will have paper back-ups while the electronic voting machine will not. Votes are saved in two places: in the eSlate's internal memory and on the mobile ballot box. After the election, these mobile ballot boxes are sent to county headquarters to be tallied by a software system, named "*Tally.*"

Battery/Power Attacks

The most basic vulnerabilities in eSlate voting machines involve their batteries / power systems. eSlate machines draw their power through a cable that connects them with the Judge Ballot Controller, its access control mechanism. Researchers connected a DSub-15 null terminator to the daisy-chain cable, which shorted the power and forced the eSlate machine to operate on battery power. To avoid detection, the attackers hid terminator at the end of the cable, tucked under the top panel of the machine. In a matter of hours, this led the eSlate machines to run out of battery power and die. Because the terminator can easily be hidden outside of plain site, this attack is difficult to detect. The "death" of voting machines during the 2016 election, discussed in Chapter 1, is consistent with this type of attack. (Alternatively, it could have been caused by operator negligence).

The reliance of eSlate voting machines on battery power also enables a denial of service attack. Specifically, removing the battery for 30 seconds leads the date/time of the JBC to get reset to 1996 (presumably a hard-coded date/time). Once this happens, the machine cannot work until someone figures out what's wrong and calls the election headquarters, where administrators can manually reset the date and time of the machine. This attack can be executed on-site, during an election, but might be conspicuous. More likely, it would be executed the night before the election by an insider.

Access Control Attacks

Once a voting machine is powered on, its primary access control

Figure 8: *A Hart InterCivic eSlate voting machine used in Yolo County, California*

Figure 9: *A JBC unit, used for access control in the eSlate. Unencrypted JBS protocol can be forged, allowing ghost ballots*

mechanism is the JBC. Every time a voter enters the polling location, the JBC will generate a unique access code, which a poll worker prints for the voter. When the voter enters the access code in the eSlate machine, voting begins. Unfortunately, the JBC uses a cryptographically insecure pseudorandom number generator that allows attackers to predict future access codes based on an initial sample. Once the access control mechanism is circumvented, attackers can cast an arbitrary number of ballots. Proebstel et al., for example, wrote a script that allows them to cast up to 10,000 votes.

Butler et al. finds a second way to access PIN numbers, by leveraging a bug in how the eSlate handles errors with its printer. During early voting, merely selecting "disable printer" from the eSlate menu function was enough to enable attackers to steal 10,000 pin numbers. During regular voting, however, the attackers noticed that the "disable printer" function was not included in the options menu. However, they learned to leverage a bug in the way eSlate's handles errors to get around this. When attackers requested an Access Code Report through the eSlate menu system, the "disable printer" function was re-enabled. When the attackers selected disable printer, they were once again able to generate up to 10,000 unique voter PIN numbers. The bug that Butler et al. identified soon became a theme in their research on Hart InterCivic machines: when the machine throws errors, it can be manipulated to display undocumented functionality that expands the attack surface and enables novel attacks.

Regardless of the way in which PIN numbers are stolen, using them to enter hundreds of votes at the same time is a time-consuming process. However, researchers have developed a mechanism for speeding up the process. All eSlate machines contain a tactile input jack. Security researchers understand that the signals sent through this jack can be mimicked using audio input. Therefore, attackers used a Python script to create the correct audio frequencies that would trigger the machine to cast votes. Each ballot takes 20 seconds to cast using this script. By combining stolen PIN numbers with the audio script, an attacker could cast more ballots than would be possible if he or she were attempting to do so manually (we assume a fixed time constraint, since the attacker needs to leave the voting location before being detected).

The JBC can also be circumvented with a "ghost voting" attack. Specifically, an attacker, sending fake button presses to the eSlate machine, can mimic the communication protocol used by JBC. The eSlate machine makes no distinction between these "soft" button presses and real "hard" button presses. In both cases, the operating system processes the signal as if the keyboard were being pressed. The operating system then sends the keyboard information to the eSlate program. Using a laptop, researchers connected to the JBC's parallel port to create these soft button presses and commanded the eSlate machine to vote for their favored candidates. This attack is possible because JBC and eSlate use unencrypted communication that is easily forged.

CHAPTER 3
THE VOTING MACHINE VIRUS

"Halderman jogged into an alleyway, where his source stood patiently, dressed in a charcoal colored trench coat and wielding a black canvas bag. After a few terse formalities, he handed Halderman the bag with the [voting] machine inside. Halderman never saw the man again" — *Politico*

If you want to know what it takes to hack one voting machine, the best way to do that is to read Chapter 2. It's as easy as signing up as an election volunteer and pressing incorrect buttons when citizens go to vote or using a Palm Pilot to transmit radio frequencies to a voting machine and casting illegal ballots via the remote device.

But you didn't pick up this book because you want to know how attackers will hack just *one* voting machine. You want to learn how attackers can hack our *elections.* In this chapter, we'll explore in-depth why voting machine viruses are the most effective way to achieve scale and materially impact election outcomes.

A voting machine virus is a type of malware — a malicious piece of code or software meant to "damage or disable" a computer system. Successfully implementing malware requires three discreet steps: first, the attacker must *inject* the target computer, and then the attacker must *take over* some software process, and finally the malware must *propagate* itself, via a wireless network or a local network, to nearby computers. Once propagated, malware can spread rapidly, which is why malware is colloquially referred to as a virus.

Attack #1: Insert Virus via Buffer Overflow
Research Team: Aviv et al. (2008)[43] and McDaniel et al. (2007)[44]
Target Machine: iVotronic by ES&S

In their 2008 paper, "Security Evaluation of ES&S Voting Machines and Election Management System," Aviv et al. demonstrate how attackers can use a buffer overflow attack to hack ES&S voting machines.[45, 46] Buffer overflows are one of the most basic and effective vulnerabilities used to insert malware into software programs. Here's how they work. Imagine you have a string. You define the string to be 5 characters. When the user types a five-character string into your program, you copy it into memory. Now what happens if the user types 50,000 characters into the program? It should probably fail. But many programs still accept the string, and copy it all in memory. A buffer overflow attack is a Trojan horse of sorts, where the attacker writes malicious code, which is copied into memory by the computer's firmware, which fails to recognize that instead of entering "Name" into an input field, you've entered an entire virus!

Well-written malware won't just insert itself into the "host" computer. It will also force the computer to run the code it inserted, thus *coopting* the host machine. The malware will now execute with the permissions of whatever functions it has co-opted.

Because buffer overflows have been around for decades, they are well understood. There exist toolkits and programs designed specifically to help programmers write programs that are safe from buffer overflows. Alas, voting machines — and in this case, the iVotronic model specifically, are riddled with buffer overflow vulnerabilities. Researchers on the EVEREST Team, which audited voting machines for the State of Ohio, demonstrated this experimentally, showing that they could insert malware into a voting machine, via its PEB access control device, and have the virus spread from the voting machine to the back-end server, called Unity, which tallies all the votes for a particular county.

Once inside the Unity back-end system, the malware ran a series of commands that enabled it to create a new account. The username was "attacker" and the password was "weonwnyou." To test if their attack was successful, the EVEREST researchers attempted to log-in directly to Unity (via the "front door," so to speak) using the credentials their virus had created: username "attacker," password "weownyou." The log-in was successful, demonstrating that their malware had penetrated the voting machine, taken over its software, and propagated itself to Unity, the tallying service.

Why is this so significant and dangerous? Well, to start with, by infecting Unity with a virus, the attackers demonstrated that they had the power to change all the votes in the entire county, not only on a single voting machine. Separately, the attackers showed that Unity operators were browsing the Internet during their downtime. This is doubly significant — if multiple Unity operators were on the Internet at the same time, the virus could spread from one county to another, creating the largest scale attack possible and creating a statewide or nationwide problem.

The success of the EVEREST attacks indicate a need to update our threat model: a single attacker can now manipulate the results not only for his/her voting machine but also for his/her entire county — and even multiple counties.

In their report, the EVEREST team identified numerous buffer overflow vulnerabilities in the iVotronic machines they studied, including:

1. Poll Opening Overflow: During the poll opening process, the iVotronic voting machine copies a string from external memory to its internal program stack. The EVEREST team used this overflow to take complete control of an iVotronic machine, including the ability to erase or change election results, and print arbitrary ballots to the real-time audit log, a printer used to verify election results.

2. "Hotspot" Image Processing Exploit: Hotspot image files are used to map parts of the touch screen to actual functionality. The source code for the iVotronic machine copies bytes into these image headers. If a malicious ballot header is written on the Compact Flash card, which manages this process, it can load a virus on the iVotronic machine. Again, this allows hackers to take complete control of the machine.

3. Supervisor Initialization Overflow: Supervisor iVotronics are access control devices used by election supervisors to configure voting machines before an election. During the initialization process, supervisor iVotronics are linked to Unity, the back-end tallying server. EVEREST researchers found a buffer overflow in the initialization process. Their attack consists of spoofing the Unity server and then connecting the supervisor iVotronic device (called a PEB) to the attacker's computer. During the initialization process, the attacker can load malware to the PEB, and thus the iVotronic device.

Figure 10: *Reprogramming the Compact Flash Chip for an iVotronic would allow an attacker to re-write firmware. As featured on ES&S's website.*

Figure 11: *An actual broken seal from a Wisconsin voting machine used in 2016. As featured on Fox News.*

In summary, the most direct way to insert a voting machine virus into a voting machine is to find a buffer overflow vulnerability and use it to insert a virus onto a single voting machine. Because all voting machines in each county are connected to a back-end server, which tallies countywide votes, the virus can potentially infect all voting machines in a single county. The EVEREST research team demonstrated this attack on iVotronic machines, which are manufactured by ES&S. This type of attack could potentially work on a voting machine of any brand or build.

Attack #2: Firmware Modification in iVotronic DRE
Research Team: Aviv et al. (2008)[47] and McDaniel et al. (2007)[48]
Target Machine: iVotronic by ES&S

Firmware is the software that is pre-loaded onto a voting machine and is written by the voting machine manufacturer. There are vulnerabilities in voting machines that allow hackers to directly modify this software by taking advantage of flaws in the voting machine's user interface. Firmware modification attacks generally (but not always) require direct access to a voting machine's motherboard, where firmware is stored. Thus, the main attacker we are concerned with here is the "insider" — a malicious election volunteer. As we have seen, insiders are rarely vetted and pose unique threats to election security.

Researchers have used the firmware overwrite framework to attack voting machines made by all four major manufacturers. We'll start by looking at a few vulnerabilities in the iVotronic model, studied by the EVEREST research team.

1. Internal Flash Chip: Using a Torx security screwdriver, an attacker could gain access to the internal flash chip socket. The attacker could then reprogram the flash chip using a flash writer program. The flash writer can be used to infect the machine with arbitrary firmware. After the initially invasive attack, this would be difficult to detect.[49]

2. Update Menu: All iVotronic voting machines come with direct access to the firmware via the firmware update menu. This menu is protected by a Quality Assurance PEB, a special access control device meant to prevent firmware updates by unrecognized parties. Unfortunately, the Quality Assurance PEB can be emulated with a simple palm pilot device, or any device that communicates using IR protocol. The only difference between the "quality assurance" PEB and the regular PEB access control device is a single character is the device's communication protocol. Once the attacker has bypassed this defense mechanism, he or she can access a firmware update menu on the screen of the voting machine. Using the CF card interface (which is external to the machine), the attacker can load new firmware onto the voting machine.

It is often the case that the CF card interface is protected by a seal. In this case, the housing needs to be removed. The attack cannot be implemented in broad daylight, but could easily be executed from the warehouse, or by a volunteer the night before the election. If the attacker chooses not to remove the casing, and the broken seal is noticed, this will function as a denial of service attack. All the votes on that voting machine will be discarded. If the attack were targeted at a district with known political preferences (i.e. Philadelphia, New York, Detroit), even a denial of service attack could materially impact election outcomes.

3. PEB Interface: The PEB interface is a slot on the front of the iVotronic machine. It is used to recognize the PEB access control device during voting. The simplest way to infect firmware while the polls are open is to use the PEB slot. Everyone has access to the PEB slot when voting. Using a spoofed PEB, the attacker can load malware onto the voting machine via the PEB interface, which reads IR protocol from palm pilots. The malware could be specifically coded to re-write firmware (or directly infect the machine). Because the PEB interface is exposed, it is a dangerous attack vector.

In all three cases, the practical implication of this attack would be to allow attackers to change votes and hence election outcomes. This attack can potentially persist across election cycles. iVotronic voting machines are rarely audited, so it is plausible that the infected firmware will remain on the iVotronic system indefinitely. According to the EVEREST report, ES&S has no way to audit its own firmware to guarantee its integrity, so this means that persistently corrupted firmware could go undetected.

One of the key defenses against firmware attacks are error checking functions in the firmware. But if an attacker replaces the firmware, they are not going to include the integrity checks in the modified firmware. Thus, the integrity checks won't prove very useful. Likewise, paper ballots are often used to verify that the electronic tallies on voting machines are correct. However, since firmware controls both the electronic vote counters AND the printers, we can assume that detecting an attack would be very difficult (and a recount would be useless). Even if firmware were to be discovered, the only possible way to fix the votes would be a redo of the election.

Evidence from the 2016 election may be consistent with one of these attacks. Several broken seals in Wisconsin precincts, which were reported to the Elections Commission, ultimately contributed to demands for a recount. A Wisconsin Elections Commission spokesperson reported that "The seals were broken by the technician who came to install the modem in the machines after purchase. The technician was supposed to replace the seals when he was finished but neglected to do so." That the voting machines seals were broken in the regular course of business — and that the broken seals were not discovered until after the election — is evidence that uninterrupted access to the chips that run iVotronic firmware is not only possible but is routine. A sophisticated attacker could make malicious changes and that attack might not be detected until after the election, if at all. [50]

Attack #3: Firmware Replacement in the AVC Advantage

Research Team: Appel et al. 2009[51]

Target Machine: AVC Advantage by Sequoia

Let's return to the AVC Advantage voting machine model studied extensively by Appel et al., as requested by New Jersey's Superior Court. Earlier, we discussed how the machine's poorly designed user interface can randomly disenfranchise voters by accident or be co-opted to systematically disenfranchise certain groups of voters. Here, we focus on how poorly designed software enables attackers to take control of the AVC Advantage and manipulate election results, using techniques such as firmware replacement and viral propagation.

Firmware Replacement: The most straightforward attack against the AVC Advantage is to simply replace its firmware. This attack can be carried out by anyone who has physical access to voting machines (for approximately 7 minutes). Princeton Professor Ed Felton has showed that gaining access to voting machines before an election is trivial (as demonstrated in **Figure 12,** from his blog). Appel's group wrote new firmware on a ROM chip. They simply copied the source code on the resident ROM chip and added 122 lines of new source code (600 bytes of machine code). Separately, Halderman and Feldman showed that you don't even need access to the original ROM chip. They showed that they could reverse engineer the AVC Advantage's resident firmware is a matter of months.[52]

With the reprogrammed ROM chip in hand, an attacker needs physical access to the AVC Advantage. Then, he or she must open its

Figure 12: *AVC voting machines are often left unattended, a vulnerability that enables firmware replacement*

Figure 13: *The AVC Advantage voting machine studied by Checkoway et al. and Appel et al.*

Figure 14: *The motherboard of an AVC Advantage. Election software is programmed directly on removable ROM chips*

casing. This can be accomplished in one of two ways. Either, the attacker can duplicate the key to the lock (the same lock is often used across all machines) or he or she can pick the lock. Appel's team did both. First, they successfully copied the key at their local hardware store. Second, Appel showed that with 1-2 days practice a novice (someone who has never picked a lock before) could pick the wafer-tumbler key lock in 15 seconds. The final step involves removing 10 screws, replacing the existing ROM chip with the corrupted chip, and then screwing the 10 screws back in. Timing this attack, Appel concluded that it could be achieved in 7 minutes with moderate practice.

This attack is dangerous because it has the potential to scale. One of the main reasons that public policy officials argue that elections are safe is because systematically attacking an entire state's election would require significant manpower. An even larger network of attackers would be necessary nationally. Here we disprove this assumption. In the factory where AVC Advantage machines are programmed before an election, an attacker could infect thousands of machines with a virus. In the warehouse, where these machines are stored, hundreds of machines are exposed (on average, 500). Finally, in a single voting location (which aggregates two or more precincts, usually), a dozen AVC Advantage machines are vulnerable to attack.

To put this attack in perspective, hacking one warehouse and shifting 20% of the vote from one party to another would change the statewide vote count by approximately 1%. In a congressional or legislative district, hacking one warehouse could change the outcome of the election by as much as 20% (though attackers would likely reduce the extent of their manipulation to avoid detection). To avoid being identified in pre-election testing, the malicious firmware is programmed to 1) only cheat in official elections, 2) only start cheating after at least 20 votes have been cast, and 3) alter the ballot counts in both internal memory and on the results cartridge (which is used to ultimately tally up the votes cast). Since internal memory is used in auditing these machines, an audit would not be able to identify this attack.

Sequoia claims that this attack is blocked by cryptographic countermeasures. Specifically, they use a checksum to ensure that no code has been added to the ROM chip. A checksum is a digit that represents the sum of the correct digits (i.e. bytes) in code. However, since attackers are re-writing the firmware entirely, this can be circumvented by either 1) ensuring that the sum of the bytes in the new program is identical to the old program (which is what Appel et al. decided to do) or 2) modifying the firmware to eliminate the check. It is important to note that cryptographic defenses don't work against an attacker who has control of the system and can re-write or eliminate the defenses. This is like building a sophisticated lock and then leaving the door open.

Viral Propagation via Daughterboard Attacks: AVC Advantage machines often come with a "daughterboard" which allows audio voting for the blind or partially blind. The daughterboard stores firmware on flash memory (rather than a ROM chip). Therefore, attacking the daughterboard requires even less effort than attacking the motherboard; the attacker must simply inject a (corrupted) audio-ballot cartridge into the machine. Audio-ballot cartridges are unique because they can store both ballot definitions and firmware at the same time. Audio ballots are inserted into AVC machines during an ordinary election, when the poll worker initiates the beginning of voting. Therefore, the attack could be carried out inconspicuously.

Since the daughterboard sends its results directly to WinEDS (the central counting system, analogous to Unity), the modified firmware can send a virus to the central system, which can then propagate it back to all of the AVC machines that communicate with it. This would allow an attacker to target an entire county with one chip. The virus is easily transferred because at the end of an election, the cartridges are sent to the cen tral county office, where the WinEDS system loads their results and tallies them. On WinEDS machines in Union County (where Appel conducted his tests) and separately in California, which conducted its own review of WinEDS, auto-run was enabled on the machines. This means a cartridge carrying a virus will automatically run the virus in the background (unnoticed) on the target WinEDS machine. At the very least, the virus can copy itself onto all cartridges that are later inserted (i.e. countywide cartridges). If the WinEDS machine is connected to other WinEDS machines on its network, the virus could spread between counties.

DOS Attack via Daughterboard: An attacker can use the daughterboard to stage a Denial of Service (DOS) attack. The AVC motherboard will throw an error if the daughterboard is misconfigured. And since the daughterboard includes date information, it can be programmed to fail on Election Day. Each AVC machine comes with only 30 backup paper ballots. This means that a DOS attack could influence voter turnout. Even if audio-voting is disabled (which is illegal), the daughterboard can still inject the motherboard with its virus (via a buffer overflow attach) and thereby guarantee denial of service.

Viral Propagation via the Internet: Basic security protocols call for minimizing the attack surface on mission-critical machines. WinEDS is certainly mission-critical; it ultimately tallies countywide votes, which are used in real-time to determine the winner of an election. Nonetheless, Appel et al. found that the government employee who operated it had used Union County's WinEDS machine for personal Internet surfing. Because the WinEDS computer is configured assuming that it will not be used for web surfing, numerous ports are left open, providing a dangerous attack surface. Using spear phishing or other social engineering techniques, an attacker may be able to infect a WinEDS machine directly via its operator. As in the case of the daughterboard attack, once the WinEDS machine is infected with malware, the virus can copy itself onto all the cartridges that are later inserted.

Attack #4: Return Oriented Programming Attacks on the AVC Advantage

Research Team: Checkoway et al. (2009)[53]
Target Machine: AVC Advantage by Sequoia

Return oriented programming is an attack to generate malware. Like with buffer overflows, return-oriented programing can create arbitrary (and often malicious) programs on the target machine by exploiting vulnerabilities in the source code. Unlike buffer overflows, however, return oriented programming does not involve writing new code. Instead, the attacker pieces together calls to existing lines of code stored in memory to create a malicious program. In C, for example, the attacker can leverage calls to libc, the C standard library, to run malicious code that has permission to access protected memory with root permissions (this class of attacks is called *return-to-libc*). Buffer overflows are often used in conjunction with return-oriented programming to gain access to the program stack.

To attack an AVC Advantage machine with return-oriented programming, one must start by identifying "gadgets" which are the short snippets of code used in the attack. Remarkably, Checkoway et al. identified a Turing-complete set of gadgets in the AVC Advantage firmware. This means they can write any arbitrary program (technically, any program that can be executed on a single tape Turing Machine). This is a feat in its own

Mnemonic	Description
`.ascii "str"`	Inserts the bytes for `str`
`.asciiz "str"`	`.ascii "str"; .byte 0`
`.byte b,...`	Insert a byte for each argument
`.data n`	Inserts n `NUL` bytes
`.var A,n`	Define a new 16bit variable A at location n
`.word n,...`	Insert a word for each argument
`label:`	Define a new label
`add A,B,C`	$A \leftarrow B + C$
`addi A,B,n`	$A \leftarrow B + n$
`and A,B,C`	$A \leftarrow B \& C$
`b label`	Branch to `label`
`btr A,label`	Branch to `label` if A is true
`bfa A,label`	Branch to `label` if A is false
`call SP,label`	Push address of the next gadget to stack at SP, jump to `label`
`cpl A,B`	$A \leftarrow \sim B$
`dec A`	$A \leftarrow A - 1$
`di`	Disable interrupts
`ei`	Enable interrupts
`halt`	Halt the machine
`in A,C`	`in a,(c)`[1]
`inc A`	$A \leftarrow A + 1$
`jr A`	Jump to address A

Mnemonic	Description
`la A,B`	Set A to the address of B
`la A,label`	Set A to the address at `label`
`ld A,n(B)`	$A \leftarrow (B + n)$
`ldx A,B,C`	$A \leftarrow (B + C)$
`li A,n`	$A \leftarrow n$
`mov A,B`	$A \leftarrow B$
`mul A,B,C`	$A \leftarrow B \times C$
`neg A,B`	$A \leftarrow -B$
`nop`	Do nothing
`or A,B,C`	$A \leftarrow B \mid C$
`out C,A`	`out (c),a`[1]
`pet`	Pet the watchdog timer[1]
`pop SP,A`	$A \leftarrow (SP)$; $SP \leftarrow SP + 2$
`push SP,A`	$SP \leftarrow SP - 2$; $(SP) \leftarrow A$
`pushi SP,n`	$SP \leftarrow SP - 2$; $(SP) \leftarrow n$
`ret SP`	Pop from stack at SP, jump to value
`seq A,B,C`	$A \leftarrow B = C$
`slt A,B,C`	$A \leftarrow B < C$
`slti A,B,n`	$A \leftarrow B < n$
`sne A,B,C`	$A \leftarrow B \neq C$
`srl A,B,s`	$A \leftarrow B \gg s$
`st A,n(B)`	$(B + n) \leftarrow A$
`stx A,B,C`	$(B + C) \leftarrow A$
`sub A,B,C`	$A \leftarrow B - C$

Figure 15: *The pseudo-assembly language developed by Checkoway et at. The language is Turing-complete.*

```
qsort: # void qsort( array, left, right )
    push array
    push left
    push right
    pet
    ld left,10(sp)
    ld right,12(sp)
    slt templ,left,right
    bfa templ,cleanup
    ld array,8(sp)
    ldx pivot,array,right
    mov index,left
    mov i,left
loop:
    slt templ,i,right
    bfa templ,break
    ldx templ,array,i
    slt temp2,pivot,templ
    btr temp2,continue
    ldx temp2,array,index
    stx temp2,array,i
    stx templ,array,index
    addi index,index,2
```

Figure 16: *A code snippet from Checkoway et al.'s Quicksort implementation, using return-oriented programming*

right. The next step is to identify a mechanism for gaining access to the stack and triggering the execution of a return-oriented payload. Checkoway et al. exploit a stack buffer overflow in a memory cartridge. For our purposes, any buffer overflow that gave us access to the stack would do the trick.

Now comes writing the actual exploit, which involves combining firmware (existing) with new return-oriented code and using a buffer overflow to trigger it. This attack can reuse all of the AVC's actual code to simulate the open/close polls functionality as well as voting behavior. In Checkoway's specific implementation, half of the votes cast for Candidate A are instead cast for Candidate B. All of the Cast Vote Records (CVRs) are changed to reflect the new votes. Because AVC Advantage does not have a paper trail (a key vulnerability, even compared with iVotronic) these changes will not be detectable, except potentially in audit logs. Finally, control is handed back to the regular post-election code, which prints the election results (now falsified) and submits the final tally.

Detecting this attack is possible using forensic techniques. However, Checkoway notes that even these forensic tests can be circumvented if the malware deletes itself from memory before returning control to the regular program. Furthermore, this attack requires only 3.2KB of RAM space, so there is enough remaining space to write more sophisticated program on the AVC Advantage. For example, researchers in Appel's group at Princeton programmed an AVC machine to allow them to play Pac-Man, among other hacks.

Attack #5: Code Insertion and Denial of Service on AccuVote-TS
Research Team: Feldman et al. (2006)[54]
Target Machine: AccuVote-TS by Diebold

Alex Halderman, who received his PhD at Princeton under Andrew Appel and is now a Professor at the University of Michigan once told Politico, "There's a lot of cloak and dagger in election security."[55] Apparently, he was referring to a private voting machine he received from a disgruntled employee in a back alley behind a hotel in New York City. "Halderman jogged into an alleyway, where his source stood patiently, dressed in a charcoal colored trench coat and wielding a black canvas bag. After a few terse formalities, he handed Halderman the bag with the machine inside. Halderman never saw the man again," Politico reported.

This odd exchange culminated in the publication of "Security Analysis of the Diebold AccuVote-TS Voting Machine" which argues that the Diebold AccuVote model is not suitable for use in U.S. elections. This was a coup for Feldman and Halderman, who had been interest-

Figure 17: *Diebold is the most computer-like among its peers. This increases the number of available attack vectors.*

Figure 18: *Volenteers read printout from Diebold machine in Lancaster, Ohio*

ed in the Diebold machine ever since a Diebold employee mistakenly posted the machine's source code (40,000 files worth) to the Internet. Rubin and Wallach, using this source code as early as 2004, showed how the Diebold could be hacked. Now Feldman and Halderman had the opportunity to test these hypotheses experimentally on an actual voting machine. There's an additional reason Diebold is particularly interesting for us: Diebold is one of the largest players in the voting machine market. At the time, its 33,000 machines were used by 10% of the electorate.

The hardware of a Diebold machine resembles a PC. It has a keyboard port, removable flash memory, 32 MB of RAM, and a 133 MHz Sh-3 RISC processor, among other features. Since computers are extremely complex, they are vulnerable to targeted attacks. Hence, there are several attack vectors than can used to insert malicious code:

Take Over on Boot: The boot process for a Diebold machine involves jumping to address 0xA0000000 in memory and booting the operating system. However, if a file called fboot.nb0 is present, that file will be used as a replacement bootloader. If the file nk.bin is present, the machine will assume this is a replacement operating system. These keywords allow an attacker to replace the operating system with a malicious copy.

File System: The Diebold machine runs a FAT or FAT32 file system, which is stored on RAM (rather than disk). This means that every time the computer is reloaded, the File System is erased. As a result, an attacker can store malicious code on the file system for an attack, and it will be deleted when the machine is shut down. This would prevent a forensic evaluation from discovering the attack.

BallotStation: Since Diebold is basically just a computer, all election systems run on an application called BallotStation. BallotStation uses Smart Cards to enable voting. When a voter walks into the station he or she receives a Smart Card, which is activated and inserted into the Diebold machine. When s/he is finished, the voter returns the Smart Card (which has been deactivated), to the polling volunteers. Later, an "Ender Card" is inserted to signify the end of voting.

Accumulator Mode: A Diebold machine can be changed to accumulator mode to allow it to tally results from its neighboring machines. The accumulator machine is then used to transmit precinct totals to the central tabular. The accumulator system is used because it saves time and reduces the amount of work necessary to transmit election results from each precinct to the central accumulator.

These elements of voting machine design enable the two high-level types of attacks:

Vote Stealing: A vote stealing attack will transfer votes from one candidate to another, while keeping the total vote tallies the same, to avoid raising suspicion. Ordinarily, this would be difficult, but the encryption key is stored in the voting machines' main memory, so the vote logs can be accessed trivially. Each vote is maintained in two places — both on internal flash memory and on a removable memory card. Both need to be altered for this attack to be successful. Vote stealing software could be injected into the BallotStation binary file, the operating system / boot loader, or a separate program (from BallotStation) that runs simultaneously (and invisibly).

Denial of Service (DOS): There are three types of DOS attacks we consider. First, the machines could be set to crash at a certain time. Second, the voting machines could be programmed to erase their entire memory at the end of a voting day. Or third, the voting machines could be programmed to alter the vote tally in an obvious and detectable way that would lead election officials to disqualify all the votes cast on the machine. In all three cases, a DOS attack has the potential to disenfranchise targeted voters.

Both above attacks assume that you can somehow inject code into the voting machines (to run your malicious programs). This is possible either by directly installing the malicious code or by creating a voting machine virus.

Direct Installation: Direct installation could be achieved in one of three ways. First, an attacker could program an EPROM chip and install it onto the motherboard. This attack requires the attacker to physically remove the Diebold casing to access the motherboard. The second mechanism for direct installation involves creating a memory card with a file called explorer.glb. This will prompt the system to run the explorer, from which attack files can be copied to the machine (or executed from the memory card). Finally, the attacker can create an alternative bootloader to load malicious software. The Diebold machine only checks that fboot.nb0 exists before running it; it does not check the files integrity. The same style of attack would work if the attacker loaded a malicious operating system, called nk.bin. Both of these second two attacks require access to the memory card slot, which is behind a locked door. Fortunately for the attacker, however, the key to that door is generic and can be purchased on the Internet. It can also be picked in less than 10 seconds.

Voting Machine Virus: Voting machine viruses are especially potent for the Diebold voting machine model because of the accumulation functionality. This means that all the memory cards will pass through the same machine at the end of the voting day. Therefore, one infected machine can spread the virus to all the other machines via their memory cards. The virus also has an opportunity to spread during the election setup. Often, ballot definitions are stored on stacks of memory cards, which are passed between machines. Memory cards are also used to update the machine operating system. Properly designed malware will pretend to execute these updates. Therefore, malware can spread if the same memory card is used to update multiple machines.

These attacks were verified to work experimentally. Feldman et al. designed their malware to alter the internal and external vote counts to prevent their attack from being detected. They specifically installed viruses via fboot.nb0. To prevent the malware from being deleted (since the FAT is stored in RAM as opposed to disk), the malware overwrites the PowerDown() function to put the machine to sleep. It also overwrites the PowerOn() function to simulate machine reboot (even though it is actually just getting woken up).

Attack #6: Firmware Replacement on Hart InterCivic eSlate
Research Team: Butler et al. (2008)[56]
Target Machine: eSlate by Hart InterCivic

As we have seen, firmware replacement is one of the most powerful attacks against electronic voting machines and is possible because digital signatures are not used to guarantee firmware integrity. Firmware replacement is easily achieved on the eSlate machine. In just 2 minutes and using only a screwdriver, Butler et al. were able to replace the internal flash memory chip, which contains the eSlate executable and configuration files. This process requires removing tamper evident seals — and is therefore detectable — but if detected, due to state laws, would function as a denial of service attack (most states simply disqualify ballots from tampered machines because voter privacy makes it impossible to determine which voters were affected). Infection of the machine with a voting machine virus or malware could be used to steal votes from voters on the given machine, and more potently, infect other machines (which will not be visibly tampered with).

eSlate firmware is surprisingly easy to modify, even if attackers prefer not use a firmware replacement strategy. The eSlate's Ethernet port — which is exposed to voters — allows unencrypted communication between the machine and any external computing device. When researchers downloaded the configuration file, they found a function

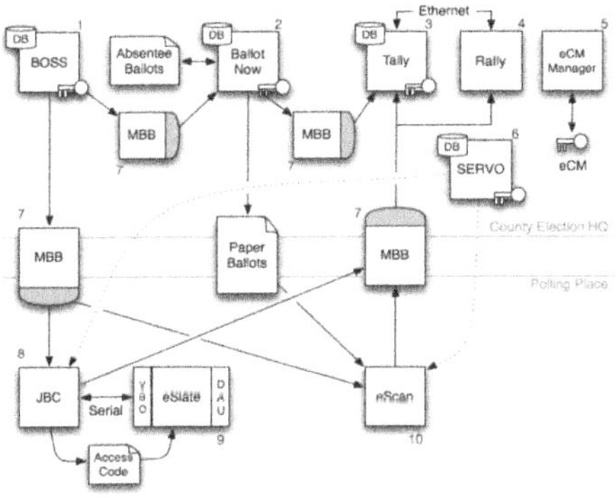

Figure 19: *System overview of eSlate system. Interaction between components increases attack surface and leads to serious vulnerabilities.*

to allow duplicate voting, which was simply commented out of the code. By uncommenting this function, the machine allowed duplicate voting. This poses a problem if voters can access multiple physical ballots (insiders) or if an attacker arrives in the morning, receives a paper ballot, takes it home and makes photocopies, and then his accomplice uses these photocopies to cast hundreds of votes later in the day. It's important to note that the audit log, which keeps track of the serial number for each ballot, can easily detect this attack. Unfortunately, it is easy to also delete the audit log.

Audit logs for the Hart InterCivic machines can be easily manipulated to destroy evidence of an attack. This is possible because audit logs are stored in a database with an exposed password (i.e. an outsider could extract the password by merely reading configuration files). Using the password and a free software package that allowed the attackers to issue SQL commands to the database, attackers were able to make arbitrary modifications to the audit logs. Experimentally, attackers modified the total vote count directly using Tally and then deleted the evidence by modifying the audit log.

Likewise, modifying the printed logs at each voting machine

(known as the VVPAT) is not difficult. An attacker with a smartphone, tablet, or laptop can plug their device in to the machine's exposed port to print arbitrary information (such as additional votes). Unplugging (or more effectively, cutting) either of the cables connecting the VVPAT printers to the voting machines would disable them. An error code will be sent to the computer device managing the election (called the JBC), but only after 15 seconds have passed. This gives the attacker time to exit the voting location before this has been noticed.

CHAPTER 4
DATA OVERWRITTEN!

"We read the code, and found really, really bad problems... Actually, let me change that, we found unacceptable problems."
— Dr. Daniel Wallach

In Chapter 3, we explained how firmware and malware attacks give sophisticated hackers the opportunity to scale up their attacks from a single voting machine to an entire county. Rather than infecting a single voting machine, hackers can write voting machine viruses to alter the vote counts for an entire county or even multiple counties. Thus, hackers can materially impact election outcomes.

In this chapter, we'll focus on data vulnerabilities: specifically, vulnerabilities that allow hackers to alter voting machine memory and thus steal votes. We place a heavy emphasis on weakly implemented encryption, which allows attackers to circumvent voting machine defenses and alter the data to achieve their political agendas. Data overwrite attacks are relatively unsophisticated, as we shall see, and can be executed with homebrew devices, potentially as small as a matchbox.

A sophisticated state actor could develop and distribute these devices, allowing them to reach dozens or hundreds of voting locations and modify the results, to favor a specific candidate or party. On a high-level, the power of data overwrite attacks is that they are technologically unsophisticated, can rapidly modify votes, and are human-scalable, meaning that with enough manpower, a small group of hackers could materially impact the outcome of an election.

Attack #1: Overwrite Unprotected Memory on M100 Optical Scan Machines

Research Team: Aviv et al. (2008)[57] and McDaniel et al. (2007)[58]
Target Machine: M100 Optical Scan by ES&S

ES&S M100 Optical Scan voting machines are paper ballot tabulators. 30,000 M100 Optical Scan machines are used to count votes in 43 states. Due to their design, the attack surface for these machines is smaller than that of touch screen voting systems. Optical Scan machines do not have computer-like interfaces. Instead, they merely scan and tabulate paper ballots. Since there is no user interface, M100 machines are more difficult to attack than direct-electronic voting machines. That said, for a sophisticated attacker, M100 voting machines could be coopted. M100 voting machines are vulnerable on numerous fronts, ranging from malware and firmware rewrites, to physical memory attacks.

PCMCIA memory cards are hardware devices used to count votes and send the results back to Unity, the ES&S back-end tallying server. Each PCMCIA memory card corresponds to a single voting location, a relationship that is hardcoded into the software system. However, when processing PCMCIA memory cards, the back-end software Unity fails to check the number of precinct results being reported, which is a hard-coded number on the PCMCIA memory card. Therefore, if an attacker were to program the PCMCIA memory card to indicate that it was reporting results for five voting locations, Unity would read an arbitrarily long string from memory (the only upper bound on the length of the string is the size of the PCMCIA memory block allocated to a precinct and the size of the integer data type).

Therefore, if the attacker can successfully write to PCMCIA memory, he or she could potentially:

1. Report Fake Votes: Since Unity is not checking the number of voting locations being reported per PCMCIA card, the most straightforward attack would be to simply report results for a non-existent precinct. The probability of success for this attack depends on the competence of the Unity operator. A well-trained operator could detect this attack if he or she notices that the total number of voting locations reported exceeds the number of precincts that exist. Secure systems should always have redundant defenses and should never rely on operator competence as the backstop against cyber-attacks.

2. Infect Unity with Malware: Unity relies on a single, shared database for each county. To change the overall vote count in Unity, we can infect Unity with malware. Because we can trick Unity into reading our arbitrary code (thinking its processing vote counts) we can effectively take over the Unity system. By inserting arbitrary code onto a PCMCIA card, which will be processed by Unity in the ordinary course of an election, an attacker could potentially modify county results, delete the database (to destroy votes), or add votes for a specific candidate. Critically, if the malware were not discovered (for example, if the attack added a small but significant number of votes to a target precinct), then the malware could affect future elections as well. Aviv et al. refer to this as a "viral loop" where a single infection could corrupt an entire county's voting infrastructure.

3. Modify Existing Votes: Modifying a single precinct's votes is easier than faking new votes. In this case, the attacker need only modify the PCMCIA data for the existing precinct. This can be done with a simple PCMCIA reader/writer device.

The underlying assumption for these attacks is that it is possible to write to PCMCIA memory. That is surprisingly easy to do. The only cryptographic protection used by PCMCIA cards is a cyclic redundancy

Figure 20: *M100 Optical Scan machines tabulate paper ballots. Photos are provided by North Carolina's Cumberland County*

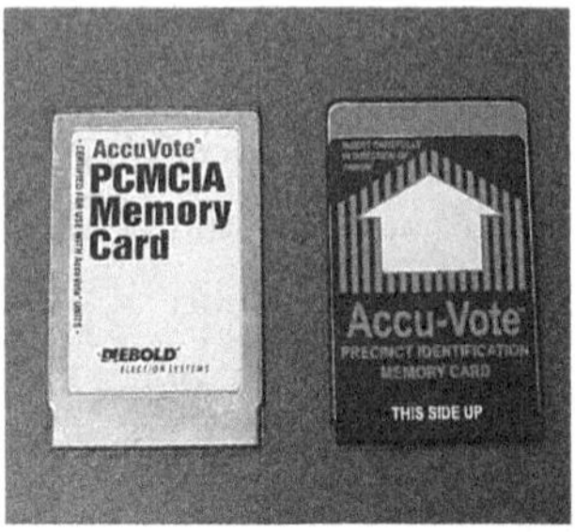

Figure 21: *PCMCIA cards are not cryptographically secure. They can be used to alter election reporting.*

check. CRC checks are error-detection codes used to identify when raw data is modified or corrupted. CRC checks are frequently used in debuggers — they are not intended to protect against planned attacks. As a result, they can be circumvented rather easily. Danny Quist outlines numerous ways to evade CRC checks in his paper "Covert Debugging: Circumventing Software Armoring Techniques."[59]

Today, the primary countermeasure against this attack is a paper ballot recount. Fortunately, M100 machines are paper ballot tabulators — they are not used to cast the vote itself. After each vote is tabulated, it is stored in a container below the M100 machine (see **Figure 20** and **Figure 21**). These paper ballots can later be retrieved for a recount. Unlike the iVotronic machine discussed earlier, forging these paper ballots would be difficult. In a limited number of attacks (i.e. when the attack is detectable) this system would guarantee the integrity of elections. However, a realistic threat model should assume that some attacks will go undetected. In this case, paper ballots don't provide complete protection for voters.

Our recommended countermeasure is strong encryption. ES&S should implement strong cryptographic protection for the PCMCIA cards. Specifically, two popular algorithms are RSA and AES. RSA is a strong public key cryptography algorithm that uses a one-time pad (large, hard to factor number) to encrypt a message and generate a public key. For example, a PCMCIA card would generate the key. Unity would need to have access to the same secret one-time pad and could use it to convert the public key into a private key and then decrypt the message using the private key. AES (Advanced Encryption Standard) is a block-cipher implementation and is used by the U.S. government. AES applies a bitwise XOR between a key and each block of 128, 192 or 256 bytes in the message text. Transposition and mixing are used over multiple rounds of key expansion to prevent uncovering the key from its repeated use (i.e. using frequency detection on the blocks). Both of RSA and AES are believed to be secure. By contrast, CRC checking — the only protection mechanism currently used by ES&S — is widely known to be insecure.

Attack #2: Cartridge Tampering in Sequoia AVC Advantage
Research Team: Appel et al. (2009) [60]
Target Machine: AVC Advantage by Sequoia

Results cartridges in the AVC Advantage can be easily modified because they lack effective cryptographic protection. Appel et al. constructed a device that allows them to modify election results after the fact and demonstrated that the WinEDS system (used to count votes in each county) fails to distinguish between real and forged votes. This "self-contained cartridge-data manipulation device" is the size of a cigarette pack and is therefore easily concealed. It can read and write data to cartridges in five seconds. A compromised poll worker could use the device when retrieving cartridges from AVC Advantage machines or when transporting the cartridges to county headquarters, where WinEDS is located.

The security failure here is that data on the results cartridge is not cryptographically protected. Appel et al.'s finding here is surprising because encrypting results is not only an obvious security feature but is explicitly guaranteed in the AVC Advantage Security Overview, which states that "after polls are closed, the Advantage immediately calculates and stores cryptographic signatures of each of the totals data files (ballot images, write in names, candidate summary totals, and selection code summary totals). The cryptographic signature values are stored in both the Audit Trail and Results Cartridge memories." In fact, the only cryptographic defenses in place are checksums (insecure) and hash functions (which are secure if used in digital signatures, but were not used for this purpose).

Digital signatures refer to the use of a secure hash function (SHA-2 is the industry standard) to guarantee the integrity of a program and/or data. Initially, when the data is created, the owner uses a secret key to sign (hash) the data. The end-user (in this case WinEDS) will confirm that the data it is receiving has been properly signed. The shared secret can either be transmitted using public key cryptography or a Diffie-Hellman key exchange, a secure key exchange protocol. Using the secret key, WinEDS should be able to authenticate the data and confirm that it was created by the AVC Advantage machine, not by an external attacker. Oddly, Sequoia does not use digital signatures in WinEDS to confirm the integrity of the votes it is receiving.

This leaves only checksums to prevent data falsification. A checksum compares the number of bytes in a data block to the number of expected bytes, to identify if that data was manipulated. Checksums are a notoriously weak security mechanism. An attacker who overwrites the election result data can also overwrite the checksum. Checksums are not meant to defend against attackers. Rather, they are supposed to be used to identify accidental errors. Finally, the type of checksum used — a simple sum of the bytes — is weak, even by checksum standards, because the attacker only needs to ensure that the total number of bytes remains the same before and after the attack (which s/he can do by adding or removing garbage bytes) to pass the cryptographic check.

Paper printouts of election results may be used to identify tampering. However, testimony in *Gusciora et al. v. Corzine et al.* indicates

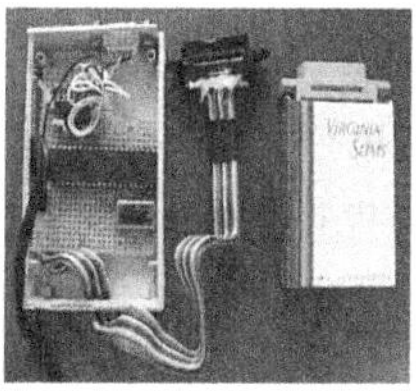

Figure 22: *A cartridge reader and writer*

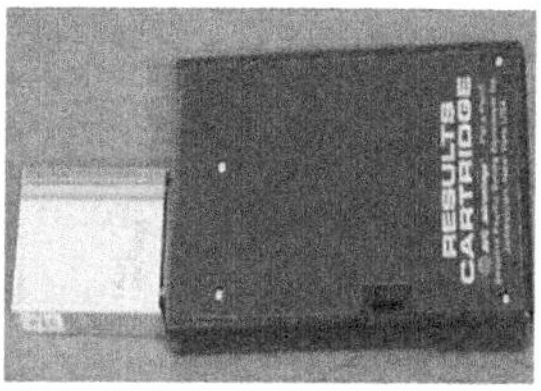

Figure 23: *A programmable device for altering votes*

Figure 24: *A matchbox sized prototype for reading and writing malware to results*

that county officials are not required to compare vote printouts against cartridge results. More concerning, precedent indicates that election officials believe the digital data should be used to resolve discrepancies, not the paper data. In Camden County New Jersey, for example, a discrepancy between the cartridge and paper printouts (thought to be the result of an electrical failure) was resolved in favor of the cartridge.

Attack #3: Cryptographic Failure in Diebold AccuVote-TS

Research Team: Kohno et al. (2004) [61]

Target Machine: AccuVote-TS by Diebold

There are several ways that tampering with Diebold machines and memory cards could affect election outcomes. Once again, these vulnerabilities are due to the lack of cryptographic protection for critical inputs, which means they can easily modified:

1. Eliminate Redundancy: Diebold uses two storage mechanisms for its system. The operating system and static system files are stored in main memory, along with backup copies of dynamic voting data. The up-to-date copy of all voting records is also stored in removable flash memory (a CD) or on a storage card. This redundancy mechanism is intended to detect improprieties. Unfortunately, this redundancy mechanism can be circumvented. Since external I/O drivers are mounted to the file system, to check for the existence of the external flash memory device, the operating system merely checks for a directory called "Storage Card." If an attacker were to create a folder in main memory called "Storage Card," he or she could direct voting data to be written to main memory instead of disk.

2. Change the Protective Counter: The Federal Election Commission (FEC) requires that all voting machines maintain a protective counter, which keeps track of the total votes cast on the machine. This way the protective counter can be compared against the total number of voters to prevent duplicate voting / election hacking. However, Diebold stores this counter in plaintext (in a mutable file). If an attacker were to change the counter, it might function as a Denial of Service attack because it would create uncertainty about the legitimacy of the votes cast. If combined with another attack — for example, using forged smart cards to cast additional votes — then it would be even more powerful. In this case, the attacker can achieve a win-win; the attacker either disenfranchises target voters (if the attack is discovered and votes are discarded), or the votes are counted, in which case the attacker is able to cast additional ballots for his/her candidate of choice.

3. Change Ballot Definitions: At the beginning of every election, Diebold operators (county election officials) must download ballots onto their machines. They have three options for doing so. They can do so manually via external input-output (I/O) devices, or via a network (either a local network or the Internet). Use of a local network would enable an attacker to modify ballot definitions through a man in the middle attack on the local network. The attacker intercepts the ballots as they are sent over the network and replaces them with maliciously designed ballots (for example, removing a disfavored candidate from the ballot). Any county election official or employee of the Internet Service Provider with access to the network could execute this attack.

4. Send Fake Election Results to the Server: Network access is particularly problematic if it is set up over the Internet. This would enable an attacker to "sniff" the connection and learn the IP address of the back-end server and then attempt to imitate a voting machine. The fake voting machine could then attempt to send votes to the server, in the hope they will be tallied. An election volunteer would have an even more direct way of executing this attack: he or she could learn the address of the backend server by checking a physical voting machine, which stores the address in plain text.

Rather than tampering with the Diebold machines, a more direct strategy for manipulating election results is to re-write the data on the memory chips. This is especially powerful if combined with a tampering attack, because it would circumvent redundancy checks by creating false vote counts both in the external memory devices and in the Diebold machine's main memory (which gets sent to the server). Here, Diebold does use some basic cryptographic techniques to protect its data, including encryption of the data and a checksum. Unfortunately, incompetent key management nullifies this defense. When Kohno studied the Diebold machine, all of the data was encrypted with a hard-coded DES key, stored in plaintext in their source code. (This key may have since changed).

Kohno et al. notes a few surprising facts about this key: 1) It is a string, not a hex value, 2) It is hardcoded, not randomly generated, and 3) This key is used on every Diebold machine manufactured since 1998 (based on the repository's commit log). Diebold's encryption was vulnerable, even without using the hard-coded key. Diebold machines used DES encryption, an encryption algorithm that has been "broken" by security researchers, meaning that it is no longer secure. An attacker could recover a DES key by 1) Using brute force, 2) Reverse engineering the key, knowing that in the case of the Diebold initialization, the CBC mode initialization vector is set to 0 (the CBC model initialization is meant to be random, and enhances the security of the encryption algorithm).

A correct implementation of CBC mode requires a random initialization vector to be created every time encryption is used. As a countermeasure, Diebold should replace DES encryption with AES encryption, which is more secure.

Diebold uses a CRC (cyclic redundancy check) to verify the integrity of the cipher text before decrypting it. However, it implements the CRC check incorrectly. Standard industry practice requires computing the checksum using a keyed hash function (like HMAC-SHA1) on the cipher text. Instead, Diebold computes an unkeyed checksum on the plain text. This may inadvertently reveal information about the plain text (especially since the CRC function is public).

Failure to correctly encrypt the voting data stored on the memory cards in voting machines means that an attacker can swap out the voting data on the memory card and write new data to the memory card, using the encryption key provided in the source code and the public CRC function to calculate a CRC value. An attacker who wanted to determine how people voted in the election would leverage the fact that Diebold stores votes sequentially. If an insider kept track of the order in which people voted, he or she could then compare those values against the decrypted votes stored on memory to link voters with their candidate preferences.

Once votes are sent to the central back-end server, votes are randomized (which means matching voters with their preferences would need to happen at each precinct, rather than by an insider at the central headquarters). However, even this defense can be circumvented. Diebold uses a linear congruential pseudorandom number generator (LCG). As with CBC mode, they do not correctly implement this technique, using static seed values for their random number generator. More importantly, LCG is not cryptographically secure, as noted by *Applied Cryptography*, which is cited in the source code's comments.[62]

Software Engineering Failures

Kohno's team attributed many of the system's failures to ineffective code design. In fact, Rubin, a co-author, later said he would have "flunked" any undergraduate student for preparing such a poorly designed system. A second factor contributing to the systems weakness were improper controls. The system was written in C++, a language that is not type safe. Even experienced programmers may accidentally introduce the potential for buffer overflow attacks when coding in C++. In their analysis of comments, Kohno's team found no indication that a control/testing mechanism was put in place to catch type-checking errors and detect buffer overflow vulnerabilities. Likewise, there was no system in place to regulate how and when developers pushed code to the central code base. This meant that a malicious developer could insert arbitrary code into the source repository and it might go unnoticed.

Kohno et al. notes that many software engineering "best practices" were ignored. For example, many critical functions lack any documentation explaining what they do. This might explain why Diebold doesn't fix many of its bugs; new developers must walk up a steep learning curve to understand the existing code base. Commit logs indicate that Diebold has no "bug database" to keep track of issues with its code, and there is no "change-control" process for monitoring bug fixes. Finally, there is ample evidence that the Diebold machines were released before the coding was complete. Comments in the code include "Okay, I don't like this one bit," "This is a bit of a hack for now" and "need to work on exception *caused by audio* I think they will currently result in double-fault."

For readers wondering how such obvious failures (such as a hard-coded encryption key!) could have made their way into live software, improper project management appears to offer a compelling explanation.

Attack #4: Mobile Ballot Box (MBB) Vulnerabilities in Hart InterCivic Voting Machines

Research Team: Butler et al. (2009)[63]

Target Machine: Hart InterCivic

In 2002, Congress passed the Help America Vote Act (HAVA) to modernize U.S. election technology after "hanging chads" created a legal fiasco in the 2000 election. Because of massive federal funding, all 50 states began a transition to electronic voting. However, almost immediately, serious issues began to arise with new direct response election voting machines (DRE's) leading the State of Ohio to commission the EVEREST report, a wide-spread study of voting machines that brought together cyber security experts from across the country. In the case of the Hart InterCivic model, Penn State researchers identified systematic vulnerabilities in the voting machines — including undocumented software features — that enabled a wide range of attacks, including vote theft and vote destruction.

Weak cryptography offers attackers the best vector with which to effect election outcomes. The most naïve attack one might consider would be to circumvent encryption entirely; copy the vote count data early in the day (by copying the bytes) and then at the end of the day, write the original data back to the device. This attack is difficult to detect because the data integrity is maintained. You've effectively destroyed votes. This is especially potent when the political preferences of a target group are well known (i.e. inner city vs. suburbs). Copying PCMCIA data — for example, using the UNIX cpio utility — does not require any password or verification. This attack was experimentally confirmed. Hart InterCivic does appear to use strong encryption, so writing arbitrary ballots to the system should not be possible.

...That is, unless you've managed to steal the encryption key. And since there is only one master key for each county, stealing the county-wide encryption key would give an attacker the ability to change election results across dozens of precedents. Surprisingly, there is undocumented feature in the Hart software system that makes it fairly easy for a malicious county election official to steal the county's encryption key and thus manipulate the election outcome. Hart InterCivic allows any election official with access to the central back-end software, called

Tally, to write the county master encryption key in plain text to the debugger file. This function can run in the background (i.e. without any notification to the user). This vulnerability is exposed via the Spyrus library, which is plugged into the election infrastructure and handles encryption. An insider could steal the encryption key and pass it along to an accomplice, who will then use it to read/write memory to the PCMCIA cards. (We showed earlier that building such a device is simple. See **Figure 17**.)

Once vote counts reach the Tally system, destroying votes is straightforward. The entire state of the election is stored in a single database, and the software is designed not to allow counting a PCMCIA memory card more than once. However, the state of each PCMCIA card is stored as a binary value in the database (i.e. 1 = complete, 0 = incomplete) An insider attack could merely flip this value to prevent certain cards (representing specific machines) from being counted. The graphic user interface for the Tally system is also modifiable by the user. Therefore, an insider attacker could modify the graphic user interface so that a future election official clicks one button (i.e. import Mobile Ballot Box) — but a completely different function gets enabled. This vulnerability is made possible because Tally used a Windows registry to monitor its graphic user interface.

The Windows registry also exposes the (again undocumented) ability to cast an "Autovote." This prints out paper ballots that are already filled in, but with a watermark that says "Autovote." Presumably, this was designed for testing. If an insider installs an image driver to the printer (as Butler et al. did in their experiment) then they can force the Ballot Now software program to print the auto-generated ballots to image files, instead of to the physical printer. This makes it easy to remove the watermark and print the ballots. An attacker then only needs to select the ballots he wants to keep (the auto-generated ballots are uniformly distributed for all the candidates) and pass them along to an accomplice, who can use them for a ballot stuffing attack. When combined with ghost voting, the duplicate ballot feature, or generating arbitrary PIN numbers, this attack becomes very difficult to detect. Uniquely, this attack would circumvent all redundancy checks, leaving only the physical sign-in books as evidence that more votes were cast than voters.

It is possible to hack into back-end software infrastructure of the Hart InterCivic system. All the passwords that control the election infrastructure are exposed to users through their configuration files. This means an attacker needs only to read the plain text configuration files, and he or she can steal the account passwords that control the software system. A second bug in the system specifies that if there are no user accounts, the first account created automatically has administrator permissions. Thus, an attacker can easily gain administrator permission by 1) Stealing an existing user's password, 2) Creating a new account, and 3) Deleting all other accounts. Then he or she can wreak havoc on the election by modifying ballot definitions before the election (denial of service) or changing the vote count after the election is complete (outcome manipulation).

In conclusion, many of the vulnerabilities in the Hart InterCivic system appear to stem from blind reliance on third party software, such as Spyrus for encryption and the Windows operating system for running the machine. Especially because the machines are very old and rarely replaced — for the most part, Hart InterCivic models in use today run Windows 2000 — the number of attack vectors have grown with time. For example, Windows random number generator function CryptGenRandom is no longer secure.

CHAPTER 5
COUNTERMEASURES

"Designing a cryptographically verifiable election is as much of a technical pursuit as it is a negotiation between scientists and the citizenry"
— Essex et al.

We know voting machines are easily hackable. Great. So far, we've focused on how to hack a voting machine and exposed the extent to which our elections are vulnerable to attack by unsophisticated attackers and state actors. The natural next question is: What can we do about it?

In this chapter, we explore the literature around safety mechanisms and countermeasures that can make our election more secure. We'll start by laying out a roadmap for how voting machine manufacturers can make voting machines less vulnerable to attack. Next, we'll consider alternatives available to the public sector, for adopting safer voting machine technologies and secure voting systems.

Roadmap for Manufacturers

Broadly, there are three primary voting machine attack vectors: 1) Circumvent access control to cast multiple votes, 2) Replace firmware with malware (and create a voting machine virus), and 3) Read/write to I/O devices to overwrite vote counts. To respond to these specific threats, voting machine manufacturers should: 1) Use cryptographically secure authorization protocols, 2) Guarantee firmware integrity, and 3) Use strong encryption to protect I/O devices.

Use Cryptographically Secure Authorization: We have discussed numerous examples of ineffective access control mechanisms, such as the PEB devices used by iVotronic, RFID-based smart cards used in Alameda County, and chip-based smart cards used in Diebold machines. Because these systems are not cryptographically protected, they are easy to duplicate or manipulate. This opens a whole host of attacks, which generally result in voters being able to cast additional ballots and/or Denial of Service. A cryptographically secure handshake is all that is necessary to protect smartcard-like devices from being duplicated. Secure ways to protect cards, include: 1) Using a secure password to identify each card and storing hashed passwords with salts on a server (not recommended, but better than the status quo), 2) Using hardware-specific keys to encrypt communication between the access control devices and the voting machine (recommended approach), 3) Use randomly generated keys and a cryptographically secure encryption mechanism (i.e. AES encryption) to encode communication between access control devices and voting machines.

Guarantee Firmware Integrity: Many of the voting machines we've discussed so far are vulnerable to outside attack because they will run arbitrary code, no matter who wrote it. This allows us to install malware on these computers, with the goal of creating a virus that will spread to other computers. One way to prevent these types of attacks is to use digital signatures to sign all software used on the voting machines. The only software updates that will be accepted by the system are those that are signed by the manufacturer. We know that it is possible to design cryptographically strong digital signatures, which would seriously impede malware developers. We might also require physical (on screen) confirmation of software updates, which would prevent technicians from unknowingly transferring malware between machines, in the case of a voting machine virus. (Feldman et al., 2006) [64]

Use Strong Encryption to Protect I/O Devices: One recurring theme in many of the voting machine models we've looked at is the use of input/output devices (i.e. removable flash memory, PCMCIA cards, PEB-devices) to store the "ground truth" information about vote counts. This means that modifying these storage devices are the fastest and easiest way to affect the results of an election. In the interest of strong security, the data on these chips should be cryptographically secure. The first step toward achieving this isto encrypt the data on these chips using a secure encryption protocol, such as AES. Additionally, digital signatures can be used to guarantee the integrity of any communication. A keyed hash function should be applied to the cipher text and stored along with the data. It should go without saying that developers should use a secure hash function, such as SHA-2, rather than an insecure hash function, such as MD-5.

Additionally, the following "enablers" are vulnerabilities that proverbially (and sometimes literally) "open the door" to attacks. They too should be corrected.

Overhaul Physical Security: The seals used on modern voting machines should be replaced with more secure seals. Trivially broken locks should be replaced with more sophisticated ones. Certainly, unique keys should be used for each lock (rather than using the same key on every voting machine, the same key that is used on many mini-bars as well). Alternately, physical access controls should be replaced entirely. Seals enable easy denial of service attacks. Simply cutting a seal would render a machine's results not credible. Furthermore, a study in Cuyahoga County, Ohio found that 15% of polling places tested had at least one seal broken. Therefore, seals don't appear to provide very effective security.

Weaknesses specified by the study included "a lack of inventory control and gaps in the chain of custody of mission critical assets." [65] Furthermore, "security seals are not checked for integrity at the end of Election Day, nor are they matched with a deployment list of security seal serial numbers. There is no attempt to reconcile memory cards intended for the precinct with the memory cards removed from the DREs at the end of the day...Therefore it is unknown whether these memory cards were tampered with during Election Day" (Feldman et al. 2006).[66]

Parallel Testing: Parallel testing involves randomly selecting voting machines (during voting) and using them for testing (instead of for voting) by casting fake ballots on them. Then, at the end of the day, you compare the votes you cast on the machines with the tallies presented by their memory cards. This is a mechanism for (randomly) testing if malware has been installed on your machines. Security experts will argue that a "secret knock" could circumvent such a test. A secret knock is a specified set of activities or actions that either turns on or turns off malware. Nonetheless, the likelihood of a successful attack falls substantially between the two scenarios. It is not only more difficult to write malware with a secret knock, but you can imagine, especially in the beginning, that parallel testing would come as a surprise to malware developers and would catch them off guard. In the long run, they may adapt and build in defenses to this system, but that doesn't preclude implementing the successful defense tactic. After all, the tit-for-tat response of attackers and defenders is a core issue in security research (Feldman et al. 2006).[67]

Software Independent Design: The most theoretically sound solution to voting machine insecurity would be to build a voting machine where any change in software can be detected in the hardware. The primary means of achieving this is by guaranteeing that every system always has a physical paper trail as a backup. This is why security researchers advocate for Optical Scan machines, where an attack on the optical scan machine is independent of the paper ballots that get stored in the "ballot box." Therefore, we can verify elections by physically counting ballots. Software independent design should be coupled with random forensic auditing of voting machines. This would enable election officials to detect evidence of improprieties and demand a recount of the paper ballots (Feldman et al. 2006).[68]

End Security through Obscurity: The source code for voting machines should be released to researchers for testing. Every voting machine's certification should require a full-fledged analysis of the software by security experts. The secrecy that surrounds election security is not useful. In fact, it is counterproductive. It is better that voting machine developers integrate the feedback of security experts than democracy get subverted when highly capable state actors decide to execute the attacks we've already outlined. In practice, the public testing of software has improved security. If we make software that it too tight for security researchers to penetrate, the likelihood of it being secure enough to protect against attackers will improve.

Always Include a Paper Trail: Most election security researchers insist that DRE voting machines include a paper audit trail, usually referred to as a VVPAT for Voter Verified Paper Audit Trail. VVPATs can be used in a recount and, since they are verified by the voter (who sees the printout after s/he votes) can be trusted. As we've noted, many of the DRE's currently manufactured either 1) Don't have a paper audit trail at all, 2) Print the results only after the election is complete (which means they may be printing false results, modified by malware during the day), or 3) Leave the VVPAT printer's cable exposed, which allows an attacker to simply unplug the printer. We recommend that every voting machine be accompanied by a VVPAT printer and that the cable be secured by plugging the cable into the back or side of the machine, which would *not* be exposed during voting.

Randomly Audit Voter Book Logs: One of the simplest ways of identifying when someone has hacked an election is comparing the number of people who physically signed in to vote with the number of votes cast. This protects against attacks where the attacker either attempts to cast too many ballots, or the attacker attempts to disenfranchise certain groups of voters. According to Dr. Andrew Appel, few counties ever look at the actual voter logs, instead relying on the Protective Counter within the voting machine. Unfortunately, the protective counter can be manipulated, or might correctly indicate the number of votes cast, without making election officials aware that some people's votes were not counted. Random audits of voter books, where individual voters physically sign their names before voting, would provide a much stronger guarantee that all voters had an opportunity to cast their ballots. Consistent with this recommendation, we've issued a request under the Freedom of Information Act to view a copy of the voter book log for Pennsylvania's District 11, Precinct 27, with the intent of comparing the number of signatures against the total vote tally reported by the FEC. So far, our Freedom of Information request has been ignored, despite repeated follow-up. (See *Personal Research* later in this book).

Create A Simple UI: We've noted that counterintuitive or poorly designed user interfaces many inadvertently disenfranchise certain groups of voters and poses an insider threat in the case of an attacker who is aware of these design flaws and intentionally takes advantage of them. Voting machine manufacturers (especially AVC) should redesign their machines to make a clear distinction whenever a vote gets cast. The VVPAT is an obvious way to do this; all voters should be informed that they can verify their vote by checking a printout next to the DRE. If the printout is incorrect, most machines have a mechanism for canceling and re-inputting the vote. Obvious bugs, like displaying "Thank you for Voting" even when the voting machine has not been activated should be eliminated as well.

Electronic Voting Machines: Alternatives and Reforms

Beyond manufacturer enabled changes to the design of existing machines, cyber security experts have proposed that we replace direct-electronic voting machines with less complex machines that will provide attackers with a smaller attack surface. The proposals include:

Only Use Optimal Scan Machines: Optical Scan Machines (like the M100) are the most secure voting machines currently on the market. Dr. Alex Halderman, Dr. Andrew Appel, and others have argued that states should replace all their DREs with optical scan voting machines. Because optical scan machines store physical ballots in a literal "ballot box," a recount will yield correct information about the voting preferences of the target group. Optical Scan Voting machines are not the perfect solution — they can still be hacked to send incorrect information to the server. If this attack is designed to be inconspicuous, a recount will never happen. Therefore, our security recommendations above — especially encrypting memory storage devices with voter data — should be implemented in conjunction with widespread adoption of Optical Scan machines.

Voter Verifiable Election Machines: Electronic voting machines are now used in 92% of counties, but because they are "black boxes" it is difficult to verify the integrity of their results. This had led to a new movement of citizens and engineers calling for a voter verifiable election system. As a general class of solution, voter verifiable protocols allow members of the public to verify the integrity of their votes. Because it's important to ensure the secrecy of ballots — both to preventing coercion, and to prevent people from selling their votes — using cryptography to make votes verifiable, but not identifiable, is central to this model. Academic research on this topic has revolved around the mathematical constructs necessary to make this solution possible.

In 2004, David Chaum and Andrew Neff independently proposed

cryptographic techniques for implementing voter verifiable elections. The so-called Chaum and Neff models have each generated a substantial body of research, analyzing the mathematical correctness of each approach, their respective shortcomings, and ways of mitigating these threats. Because there remains widespread disagreement about which implementation of secure, cryptographic voting is superior, we outline both models in our discussion below. Next, we'll review the Eperio model, which is a concrete implementation of the theory behind cryptographic voting (it's an actual software package available for download).

Visual Crypto-Receipts

Any secure voting system has two prerequisites. First, it must be impossible for the government to identify who you voted for (to avoid coercion) and second, it must be impossible for you to prove who you voted for (to avoid people selling their votes). David Chaum proposes that voters receive an encrypted receipt after they vote. Each receipt comes with a serial number and is also posted to the Internet after the election. Voters can then go online and compare the "official" copy of their vote (online) against their receipt to make sure it was tallied correctly. Chaum uses a layered receipt, where the top layer indicates the name of the candidate you voted for, and the bottom layer contains a cryptographically generated image that can be used later to verify that the vote was tallied correctly. The top layer must be disposed of before leaving the polling location, whereas the bottom layer is taken home a receipt. [69] Chaum offers a mathematical proof to explain how this is possible.

This model crowdsources integrity checking, allowing voters to have complete confidence in the correctness of their vote. Furthermore, it scales well. In an election with little suspicion of fraud, few voters will take the time to check their receipts (Chaum's estimate of 5% is reasonable). However, if fraud is suspected, a larger number of voters will check their receipts. This process of random auditing is a simple but efficient way of guaranteeing that no foul play occurred during an election. Chaum argues that an increase in people's confidence about the election's integrity will improve voter turnout. The key difference between Chaum's scheme and Neff's is the visual verification technique.

MarkPledge

Neff's encryption scheme also generates receipts. Neff argues on behalf of an encryption scheme that involves mutually distrustful trustees. Each trustee contributes a secret key to an algorithm that will then generate a master key for encryption. Decryption will require the participation of all trustees to function correctly. All votes are encrypted in cipher text, which is printed on the receipt. This is weaker form of verification than Chaum's model, because the voter can't see who he or she voted for before the receipt is printed. However, it will still confirm if there were any problems in the direct electronic voting machine. After the election, all the electronic voting machine receipts are posted to an online bulletin board and voters can compare their receipts to those posted online.[70]

Neff's and Chaum's model are fundamentally similar in that they both use receipts to empower voters to confirm that their votes were tallied correctly. The precise cryptographic tools used — and the mathematical rules behind them — offer much room for debate. The models are both theoretically complex and are unlikely to be implemented because the cryptographic tools are brand new and have not been stress tested. Deploying these models without system-wide testing could pose serious risks for election security, a topic discussed extensively by Karlof et al.

Eperio

Unlike Neff and Chaums' models, which are primarily theoretical in nature, Essex et al. have proposed a decidedly low-tech solution called Eperio. Their logic — that a simple, but secure cryptographic voting mechanism is more likely to be adopted by politicians — appears sound.

"Designing a cryptographically verifiable election is as much of a technical pursuit as it is a negotiation between scientists and the citizenry," they write. Eperio is an auditing tool that requires an auditor (any member of the general public) to download a set of spreadsheets from a central election site (i.e. FEC) and run a brief script to compare them. Eperio is an actual software package that can be readily downloaded from the Internet.[71]

Eperio has three components: a modified ballot, a receipt, and a mathematical proof that it works. The method for randomized receipts in the Eperio system involves randomly shuffling the list of candidates and then keeping only the number of the candidate selected. The list of candidates gets shredded after each person votes, ensuring secrecy. Eperio relies on 3 types of auditors (with varying degrees of engineering knowledge): the voter who must compare his/her receipt against the master data set, the election data auditor who analyzes all the data for the election, and the protocol analyst who needs to confirm that the entire protocol is secure (this requires the most technical knowledge).

As indicated in **Figure 25,** each ballot is converted into *n* entries in a table, where *n* is the number of candidate choices. The first column represents the ballot's unique id and a corresponding value for the can-

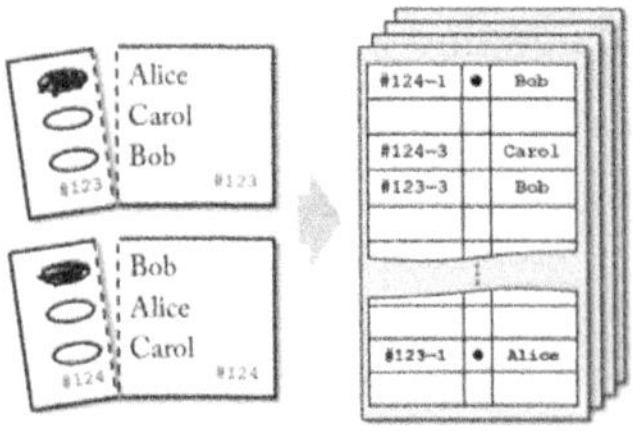

Figure 25: *Eperio's mechanism from converting voter selection cards into an Excel table that can be reviewed after the election.*

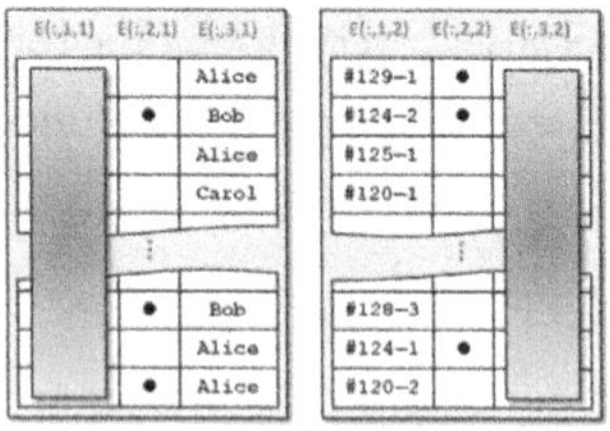

Figure 26: *Eperio as it might appear during auditing. Each table is independently and randomly shuffled.*

didate listed. The second column indicates if this choice was selected and the third column indicates which candidate this refers to. Critically, the Eperio algorithm requires that no individual can have access to all three columns. This is important because a voter with his/her receipt and access to the three columns could prove that s/he voted for a specific candidate, which violates the assumptions of End-to-End election encryption (E2E), which states that ballots must be confidential and also cannot be confirmed after the fact.

Given the data tables in **Figure 26** (right) a voter can verify that his/her vote was cast correctly by comparing his/her receipt against to the table on the right. If I filled in bubble 123-1 (a vote for Alice), and I see a dot next to that bubble, I can confirm that my preferences were indeed recorded correctly. Likewise, the election data auditor, of which there will likely be many appointed by both political parties, can count the number of dots next to each name in **Figure 26** (left), to ensure that the total vote tally for each candidate is correct. If both voters and the election data auditor find no problems, then the election was a fair election.

But how do we know that voters and election data auditors are looking at the same tables? And how can we guarantee the tables are shuffled randomly, to prevent leaking private information? Answering these questions is the responsibility of the protocol analyst. Fortunately, the entire Eperio system is open-sourced, which provides ample opportunities for researchers to identify flaws in its logic. Eperio relies on the correctness of the tables that it generates to work properly. Therefore, to increase confidence in the system, Essex et al. propose that the tables be computed on a semi-trusted computer, monitored by mutually distrustful trustees.

Open Source Voting Machines: eVACS is an open-source voting machine currently in use in Australia. The technology was developed after a close election the Australian Capital Territory, which is one of 8 Australian states, where the differences between two candidates in an election was smaller than the number of ambiguous ballots. "We'd been watching what had happened in America (in 2000), and we were wary of using proprietary software that no one was allowed to see," Australian Capital Territory election commissioner Phillip Green told *Wired.* "We were very keen for the whole process to be transparent so that everyone — particularly the political parties and the candidates, but also the world at large — could be satisfied that the software was actually doing what it was meant to be doing." [72]

For similar reasons, open source voting has become an important area of research among U.S. cybersecurity professionals. Their core argument is that by removing the "black box" of direct-recording electronic voting machines, U.S. election officials could create an environment where vulnerabilities in voting machines are rapidly resolved. This has the added benefit of creating an incentive for each political party to invest tangible resources to identify vulnerabilities in the open source voting machine, to prevent cheating by the other side. Thus, open source software enhances the incentive structure to resolve problems when we have mutually distrustful parties.

In the U.S., the EVM project is one attempt at building an open source voting machine. This project, led by Keller et al. and their Open Voting Consortium has the potential to create an alternative voting machine that could replace the standard iVotronic, Diebold, and AVC models in future U.S. elections. The Open Voting Consortium is building a voting machine where users input their choices on a direct recording electronic and then a ballot is printed out, which is machine-readable (using barcodes), and serves as the "authoritative truth" about the voter's selection. This system has two benefits. First, it allows voters to independently verify that the voting machine software is correctly tallying their vote, and second, it provides a fail-save mechanism. In the cast of a conflict, officials can manually recount ballots or mutually distrustful parties can independently count the ballots on their own computers and compare the results.[73]

According to Keller et al. printing ballots from a voting machine is preferable to optical scanning because it allows automatic conversion between languages, can be used by the blind, prevents undercounting or over-counting of votes (it guarantees one person gets one vote), and critically, eliminates the problem of extra markings or ambiguous ballots that will not get counted. Keller proposes using the AutoMark system, which is similar to a direct-recording electronic voting machine and reads the user's marked choices. The AutoMark system uses digital signatures to prevent ballot stuffing. Populex is an electronic ballot printer that is used next in the process. Because the ballots are completely uniform now, they will be easier to count (barcodes further improve the ability to count votes).

OVC does not generate voter receipts, but Keller et al. argues that its paper ballot print outs still constitute voter verified printed ballots (VVPB). This seems like a practical solution relative to printed receipts, which as we've seen, require complex cryptographic gymnastics to be feasible. Keller et al. further notes that this system is superior to the printers used in many direct-recording electronic voting machine models, because the voter can verify the correctness of their ballot, and this ballot (not an electronic record) will be used when the votes are ultimately tallied.

OVC uses cryptographically secure access control mechanisms to prevent voters from casting multiple ballots. Keller et al. argue that the specific choice of access control mechanism is less important than the cryptographic mechanisms used to protect it. Instead, they abstract away the physical mechanism as being some "token" which authorizes the user to vote. The token can be verified using a digital signature. This token could be a pre-printed ballot, poll-worker activation, a per-voter PIN, a per-precinct token or smart cards. In any case, the token is used to verify that a voter casts only one ballot. In the case of a voting error, the ballot is marked as "spoiled" and saved. The voter is then given a new token, so s/he can cast a valid ballot. This abstraction is

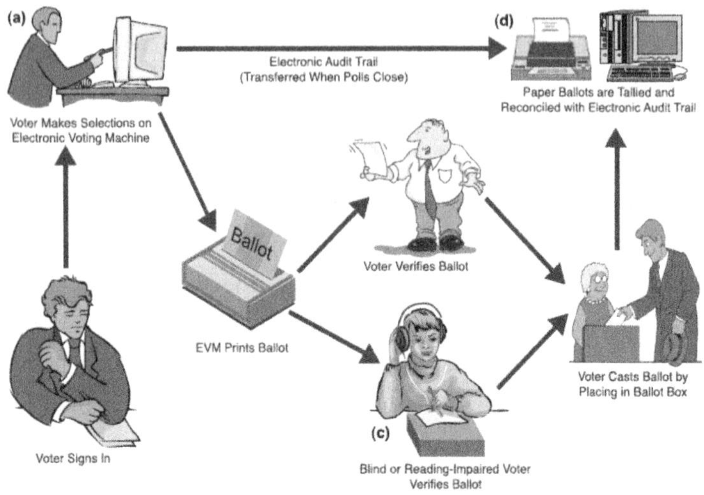

Figure 27: *A schematic of the steps involved in the OVC system*

especially useful given that most manufacturers spend their energy on the physical implementations of their system, rather than the element that matters most — ensuring that the token used by voters is unique and cannot be forged.

The OVC's physical voting machine will consist of a computer (touch screen) surrounded by privacy curtains. Installation of the computer will be an "unfold-and-plug-in" operation. Keller et al. proposes a few mechanisms to prevent tampering. First, seals and locks will be used to prevent unauthorized access to the machines. Second, the operating system and ballot definitions will be loaded from a CD rather than a hard drive (this reduces the outsider threat at the expense of increasing potential insider threats. Importantly, it prevents persistent malware from being used in multiple elections). Keller would also implement a verification process to ensure that the correct software had been loaded to the machine. A reading impaired interface would be included for the blind (audio ballots).

When a voter enters his/her preferences into the OVC, a paper ballot is printed with a barcode and a text display of the choice (human and computer readable). A Ballot Verification Station will exist so that voters (especially the blind) can verify that the barcode correctly reflects their choices. The Ballot Reconciliation Station will be used by county election officials to ensure that the physical ballots in the ballot box are aligned with the ballots saved on the voting machine's external memory. This provides redundancy in the process and will help poll workers identify any irregularities. Absentee ballots will be processed separately, using Optical Scan ballots.

The OVC system has been implemented in practice, and therefore provides a tangible and realistic solution to security dilemma posed by current direct-recording electronic voting machines in use. A demonstration of the system was conducted in 2004 at the Santa Clara County Government Building in San Jose, California. All the code for the project is readily available on SourceForge, Project EVM2003.[74] According to Forbes, an open source voting machine will be ready for use during the 2018 election,[75] after a delay caused in part by the IRS refusing to grant open source developers non-for-profit status.[76] San Francisco will be the first major city to deploy open source voting machines during its 2019 city elections. Its contract with Dominion (Sequoia) ended on January 1, 2017. This comes after a 2013 California Senate bill made it legal for cities to invest in open source voting machines and a November 2015 6-0 vote by the San Francisco Election Commission approved a move toward open source voting.

San Francisco's Board of Supervisors has said that the open source system will be cheaper than renewing its proprietary contract with Dominion. San Francisco had previously spent more than $20MM on Dominion technology, a third of which went to hardware. Since open source voting machines will use off-the-shelf parts (like PCs), the OSET Foundation, which is driving the open source movement, says it can produce voting machines at a lower cost than Dominion. Los Angeles is taking steps to be the second city to adopt open source voting. City officials say it will use open source voting machines in the 2020 election.[77]

Web/SMS Voting: Considering rampant criticism of electronic voting machine technology, a growing number of e-voting advocates are calling for use of web/SMS technology in elections. There are two frequent arguments made by non-technical political figures on behalf of e-voting: 1) If banking is secure on the Internet, then it should be possible to design a secure voting system, and 2) Online voting might increase voter turnout. We attempt to validate/invalidate these claims, looking to technical experts who have argued on behalf of e-voting.

The academic literature on e-voting is relatively sparse, which may reflect a hesitancy on the part of cyber security experts to embrace online voting (or even SMS voting). Qadah et al. proposes and develops a prototype for an online voting system in the United Arab Emirates. His team experimentally verifies the feasibility of online voting in a university setting, not going so far as to propose that the system could be adopted in a national political election. As a result, his team focuses on "how" of implementation (database management, XML management), rather than the "what" of security protocol.[78] By contrast, Myithi et al. offers an SMS-only voting system for use in India, where rapid mobile phone adoption has far outpaced Internet use.

Myithi offers no protection against duplicated voting, arguing that his system guarantees "one phone, one vote." Such a system would not be acceptable to U.S. voters.[79] Alvarez et al. and his team of researchers at CalTech, by contrast, study e-voting "in the wild," reviewing evidence from limited trials of the technology in binding public elections in Switzerland, Estonia, the United Kingdom and the United States.[80] Alvarez argues that Internet voting is more akin to absentee voting by mail than it is to the use of direct electronic voting machines. Not only does online voting expand the timespan during which citizens can vote, but it also allows remote and convenient voting. Among the nations

Figure 28: *Estonian voters can vote online, using their national ID card. Voters must buy a $7 card reader to scan their cards on a PC.*

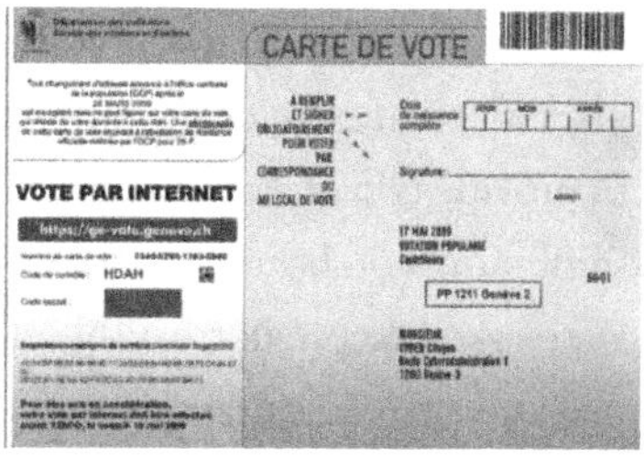

Figure 29: *In Geneva, voters receive a unique card in the mail, which they combine with a shared secret to enable voting (i.e. date of birth)*

that have experimented with online voting in real elections, Estonia, which is a full member of the European Union, has gone the farthest. In 2007, 5.4% of voters cast their ballots online, compared with 1.4% in 2005.

Evidence from Switzerland indicates that online voting may not increase voter turnout, as some non-technical experts have speculated. Alvarez reports that there is "scant evidence that Internet voting has increased turnout, though it may change the nature of participation." In Switzerland and Estonia, online voting found strong popular support and did not favor a specific political party. By contrast, experiments with online voting in the United States have been far less expansive. None of the four major trials — which include the Alaska and Arizona presidential primaries in 2000, the 2000 general election, and the 2004 Michigan Democratic presidential primary — led to any follow-up research or further deployment.

We start by turning to the problem of voter authentication. Estonia uses national identity cards with embedded digital signatures, combined with unique citizen PIN numbers (the equivalent of our social security numbers) to verify voters' identity. Alvarez et al. notes, "The lack of a nation-wide digital signature program provides an obstacle that has to be overcome in the process of authentication." Geneva developed an alternative system based on the online authentication principle that says strong authentication can be achieved by asking users to provide two of the following three things: 1) something you know, 2) something you have, and 3) something you are. In Geneva's case, voters were asked to provide a PIN and an ID number mailed to them on a unique voter ID card mailed to them in advance of the election, as well as their date of birth and city of birth. Geneva explicitly decided not to use digital signatures, arguing that these signatures are usually authenticated by Internet browsers, which can be compromised.[81]

In Estonia, after a user queries the online voting website, the voting server checks a voting registration database and then attempts to authenticate the voter. All of this communication is encrypted. If the voter is successfully authenticated, he or she "signs" a digital version of the absentee ballot using the digital signature embedded in all national ID cards. Once complete, the encrypted ballot gets sent to a server for tabulation. According to Alvarez, who was invited to observe Estonia's 2007 election proceeding by the Estonian National Election Commission, during the last 60 minutes of the election, online ballots were decrypted using a set of hardware keys, held my members of the Election Commission. Decrypted ballots were then placed on a CD-ROM, where they were counted by the election commission and KPMG, the auditor.

Statistics from this election do not indicate any preference by certain groups for online voting. There was no age or gender bias in the online voting population relative to the general population whole. However, some parties received a disproportionate share of the online vote relative to the paper vote, but not enough to sway the outcome of the election. Tech-savvy citizens were likewise overrepresented among the online voters, according to a telephone survey conducted after the election. As we might expect, online voters were 20% more likely to have used the Internet to research the candidates before voting.

Alvarez et al. makes no attempt to generalize their findings to the U.S., and, as we shall see, America may face unique threats, with its highly polarized electorate of more than 300MM voters, as compared with less than 1MM relatively homogenous voters in Estonia. Indeed, there has been no comprehensive security audit of Estonia's election. Through Alvarez characterizes it as having been successful, we can't know for certain that votes were not tampered with. We can be certain that in the U.S., where state and non-state actors have financial and political incentives to sway election outcomes, the threat model is quite different. Two additional concerns are worth mentioning. The first is privacy. Despite assurances that voters will not be linked with their votes, any comprehensive authentication strategy online will raise privacy concerns in the United States. Likewise, the ability to cast one's vote online — even if, as in Estonia, voters can change their vote at any point before the election is over — may cast doubt on whether the system is truly coercion-free and prevents people from selling their votes.

With these concerns in mind, we turn our attention to the security vulnerabilities inherent in online elections. In his article "If I Can Shop and Bank Online, Why Can't I Vote Online," David Jefferson, a security researcher and Board Member of the California Voter Foundation, argues that "computer and network security experts are virtually unanimous in pointing out that online voting is an exceedingly dangerous threat to the integrity of U.S. elections." He argues that 1) There is no way to guarantee privacy in an online election, 2) Online ballots are subject to fraud, and 3) Encryption may not provide strong protection against the NSA.

Jefferson specifically criticizes the often-made comparison to online banking, noting that financial institutions actually lose billions of dollars every year as a result of fraudulent transactions. This is despite the fact that banks deploy huge budgets to combat fraud, far more than most governments would be willing to spend. The only reason customers still believe online transactions are safe is because banks absorb the losses (which then get passed on to customers through higher prices). In fact, just as with online banking, online elections are subject to fraud.

Attackers routinely use phishing and/or malware to steal credentials from unknowing users. These types of attacks could be used to steal voter credentials and cast votes on their behalf. For example, a botnet named Zeus installed malware on people's computers and then stole people's PINs and passwords when they logged into online banking sites. Then Zeus transferred money out of people's accounts and created fake banking statements to avoid being discovered by the victims. Encryption cannot protect customers from malware, which can track keystrokes. Today, it is possible to rent a botnet for $9/hour.[82]

Critically, Johnson notes, elections are very sensitive to fraud. A 1% shift in total votes may be enough to sway local or statewide races, and if targeted at key swing states, it could affect a national election. Furthermore, unlike in online banking, there would be no paper trail in an online election, and because of ballot secrecy, there would be no way to confirm if at attack had happened. Online authentication would be particularly difficult in the U.S., where there are no national ID cards with embedded digital signatures (like Estonia) and where credentials like place of birth are unlikely to provide strong enough protection (especially if credentials can be stolen). Finally, system failure might be catastrophic. If a single person is able to achieve double voting capability, the attack might be able to scale to thousands or millions of votes. This would be a nationwide denial of service attack, because with voter privacy restrictions, they might be no way to distinguish fraudulent ballots from real ones. (By contrast, even electronic voting machine attacks are difficult to scale beyond a single county).

Online elections might be vulnerable to local network attacks and standard website attacks, known as DOM-tree attacks:

Man in the Middle: A man in the middle attack involves intercepting and manipulating communications between the victim and a server. Two man in the middle attack vectors are relevant to online voting: DNS spoofing and ARP spoofing. DNS spoofing involves intercepting communication between a client and the DNS name-server, which translates a website's name into the IP address used to contact it. DNS spoofing requires that the victim and attacker be on the same network (i.e. in a coffee shop). Computers usually cache the DNS server, which means that only one query (the initial DNS request) needs to be intercepted to initiate a man in the middle attack. A second form of man in the middle attack involves ARP spoofing on a Local Area Network. ARP protocol is the local protocol used to reach a specific computer, within an IP address. You can think of this as a building number, relative to the name of a block of houses. ARP spoofing/poisoning takes advantage of the fact that ARP protocol is not encrypted or verified. Therefore, at attacker can "poison" the client to convince it that the attacker's machine is the server. Likewise, the attacker can convince the server to associate the victim's IP address with the attacker's MAC address, the computer's unique identifier, thus impersonating the client. Once a man in the middle attack has been initiated, the following two attacks are possible.

a. Denial of Service: The attacker, having successfully initiated a man in the middle attack, can prevent packets from being sent to the target server (the election website). The attacker might hope that a discouraged voter would give up on voting through the website (as opposed to trying again elsewhere).

b. Session Hijacking: Session hijacking is a more potent attack, also enabled only when a man in the middle attack has been initiated. When the voter queries the DNS server for the election website, the attacker sends the user to a fake voting website that is designed to look exactly like the correct website. If the user ignores SSL certificate warnings (as many users do), it is easy to steal his/her credentials (because the connection is not encrypted). The attacker might reasonably rely on the fact that a substantial number of voters either: 1) Will not notice the attack, or 2) Will decide not to report the attack (because paperwork and bureaucracy is a reasonable deterrent). Using the voter's stolen credentials, the attacker can cast a vote on his/her behalf.

Cache Poisoning: Cache positing is a second mechanism for initiating a denial of service or session hijacking attack. Cache poisoning relies on the fact that clients cache the mapping between web address they query (i.e. USElection.com) and the IP address that represents the target site. Cache poisoning is the process of injecting a false mapping into the client's DNS cache, so that when they type the URL of a website they are looking to access, they actually end up at your fake website. Cache poisoning is often used in conjunction with DNS spoofing. The attacker spoofs the DNS, injects the fake mapping, and then relies on this fake mapping to hijack the user's session. The end result is that the attacker will steal your username and password for the election website, and then steal your vote.

Packet Sniffing: Encrypted connections are not necessarily protected against packet sniffing, because if the attacker can sniff the packets involved in the TSL handshake (the secure mechanism by which you access the Internet), then the attacker can defeat the encryption process. If voters use insecure Wi-Fi networks (like public Wi-Fi networks, or WEP-encrypted networks which are not secure), then packet sniffing can be used to read the data being communicated. This enables an attacker to steal credentials from the user. These credentials can be used to steal votes or de-anonymize them, enabling coercion and vote selling. Furthermore, packet sniffing can be used to enable a Man in the Middle attack (and hence, Denial of Service / Session Hijacking) because the attacker now knows how to encrypt his/her communications to mimic the target server.

The preceding attacks took advantage of vulnerabilities in the local networks used by voters to connect to the election server. They are not vulnerabilities in the actual election site. Therefore, they cannot be prevented by a well-designed election website. By contrast, the following attacks exploit vulnerabilities in the design of the target website. There do exist feasible countermeasures to protect again these attacks, which we will outline. However, as we learned from the bungled deployment of Healthcare.gov, the carelessness of government programmers should not be underestimated.

XSS Attacks: Cross-site scripting attacks (XXS) involve injecting malicious scripts into trusted websites. Usually, they involve sending the victim a phishing email (i.e. "Time to Vote! Click here to Cast Your Ballot") where the URL for the trusted website injects a script into the website that gives an attacker control over the user's interaction. XSS enables the attacker to steal user credentials while simulating correct website behavior. To defend against XSS attacks, websites need to ban users from injecting code into the website. One mechanism for doing this is escaping tags like "<" and ">" which are necessary for inserting scripts. The website can also force-sanitize inputs or create a "whitelist" for acceptable inputs.

CSRF Attacks: Cross-site request forgery (CSRF) is an attack that leverages the user's session tokens to make unwanted requests. CSRF attacks usually begin with a phishing email that sends the user to a web page that is either blank or contains decoy content. In the background, the attacker's website is running a script to steal your session tokens from a trusted website (i.e. a bank, or in this case, the election site). The script will then use the token to make unauthorized requests (such as unauthorized money transfers, or in this case, changing your vote). Remember than in Switzerland and Estonia, you can change your vote repeatedly and that this is a necessary feature to prevent coercion, vote selling, or simple mistakes. This is especially dangerous because the voter would never know that their vote was changed and cannot verify their vote (again, to prevent vote selling). Countermeasures against

CSRF attacks include: 1) Challenge tokens that can be randomized with every session and hidden within input forms. Since the attacker doesn't know the challenge token, s/he will not be able to submit requests. 2) We can reset the challenge token on every page. 3) Websites should always automatically log users out after a period of inactivity. 4) The website could guarantee the user has loaded the relevant HTML before processing the request. 5) We can verify the referrer or origin header to ensure the user is actually navigating the site (not just clicking on a URL). 6) The website can implement strong authentication with CAPTCHA, Re-authentication, or a one-time token.

SQL Injection: SQL injections are the least sophisticated website attack. SQL injection involves writing malicious code directly into a search bar or input box, assuming that the text will be used in a database lookup. In the case of an election, the injection could be used to learn how other people voted, to change their votes, or to delete the voter database. Because this is a naive attack, it is usually not possible because defenses are in place to prevent it. Defenses against SQL injection include sanitizing user input and parameterizing SQL queries (SELECT user from TABLE where user = ?). Such "prepared statements" force the user's input to be treated as a parameter and not as SQL code.

Biometric Voting: Criticism of online voting and electronic voting machines has resulted in a growing number of scholars calling for a biometric authentication system to enable e-voting. Ahmed et al. are proponents are this system, arguing that biometrics will enhance the privacy and reliability of elections. Their scheme is not complex. In the case of in-person voting via direct-recording electronic voting machine, it is effectively a way to modify the sign-in process, which currently requires photo ID. In the case of online voting, it's primary use case, it replaces national ID cards with digital signatures (Estonia) and shared-secret based authentication (Switzerland) with biometric scanning. Ahmed et al. illustrates their model with a prototype, which relies on a fingerprinting system. The fingerprinting module generates a Smart token that can be used to cast a ballot and ensures voter privacy.[83] Their system is also notable because it uses an MD5 hash function, which is not cryptographically secure.

Biometric scanning may enhance the security of online Internet voting systems by reducing the number of attack vectors discussed earlier. Token theft, coercion and bribing would still pose serious threats to this system. Another concern is with the use of biometric scanning for online voting is that it requires voters to buy a fingerprinting device, which might be expensive. This would create a socio-economic disparity in the convenience of voting and thereby create unequal outcomes for different classes of voters. More broadly, the United States lacks a central biometric database for its citizens and would likely face resistance from citizens if it tried to implement one, because of privacy concerns.

Biometric scanning could enhance authentication procedures in direct-electronic voting machines if the DREs were to be retrofitted to use fingerprinting to authenticate users (rather than a PIN or smart card). However, retrofitting the voting machines would likely require replacing them completely, which may be prohibitively expensive. Again, the U.S. does not have a central biometric database for citizens, so this would be impossible. In this case, there is the additional worry that even if a database were to be created, attackers would not register, and would therefore circumvent this defense. There is also no guarantee of enhanced security, since DREs are vulnerable to user's overwriting firmware or infecting the machines with viruses. These vulnerabilities could be used to forge biometric tokens or authenticate forged fingerprints.

In conclusion, there are many countermeasures that the U.S. government and voting machine manufacturers should implement to protect U.S. elections. In this next chapter, we'll outline our recommended approach toward tightening U.S. election security and defending the integrity of U.S. the U.S. election system, and ultimately, U.S. democracy.

CHAPTER 6
RECOMMENDATIONS AND PERSONAL RESEARCH

"The companies that produce voting machines have poured gasoline onto the smoldering embers of concern." Their "systems have vulnerabilities that 'do show incompetence" — Keller et al.

Americans no longer have faith in the integrity of U.S. elections. Just one-third of voters believe that their votes will be counted as cast. This doubt is not misplaced. The evidence has been unambiguous for a long time: in their current form, U.S. voting machines offer no guarantee of a fair election. Though coordinated attacks may be difficult to execute, a well-funded state actor could plausibly launch a targeted campaign to change the outcome of U.S. elections. Whether by denial of service or by outright vote manipulation, a coordinated attack on our election would challenge the fabric of our democracy.

One of the key flaws in our voting system is the lack of verifiability of votes cast on electronic voting machines, which are used in 92% of counties.[84] The software behind these machines is a well-kept secret and manufacturers have sued to prevent disclosure of how their software works. This "black box" — which has been penetrated by numerous cyber security researchers — offers voters little reason to be confident that their votes will be recorded as intended or counted as cast. If anything, more than a decade of research into the inner mechanisms of voting machines should make us very skeptical about the level of security voting machines provide.

The attack vectors that make voting machines vulnerable are extremely serious and very diverse. From access control vulnerabilities, to firmware replacement, to vote manipulation, to social engineering and user interface flaws, attackers have many options available to them with which to manipulate our election results. Furthermore, as Keller et al. has noted "The companies that produce voting machines have poured gasoline onto the smoldering embers of concern." Their "systems have vulnerabilities that 'do show incompetence,'" Keller continues, "And were built by 'programmers who simply don't know how to design a secure system.'"[85] The aggressive tactics used by these companies to keep their software from the public eye — combined with glaring software engineering failures — have led observers to question whether voting machine manufactures are prioritizing profitability over safety.

Technical and non-technical observers are worried that rapid implementation of digital voting has made our elections less secure, rather than more secure. In the old days of paper ballots, there were ambiguities about the intent of certain voters — hence the hanging chad debacle of 2000. But now, voters have no way to guarantee that their votes were not maliciously changed. This is not because it's impossible to guarantee security of digital system, but appears to be due, in a large part, to the incompetence of the government contractors who build these machines. For example, Diebold hard coded encryption keys in removable memory, ES&S built access control devices that could be easily simulated with a Palm Pilot, and AVC lacked digital signatures for its firmware, allowing the firmware to be completely replaced in under 7 minutes. As Wallach et al. noted, some of these mistakes are so glaring that even a novice programmer (i.e. a student) would be failed for submitting such low-quality code.

Worse, after more than a decade of research has cast serious doubt on the security of major voting machines, manufacturers have made little or no attempt to fix bugs in their machines. For example, Diebold's hardcoded encryption key never got replaced, even in future iterations of the model. When Dr. Appel provided expert testimony in a lawsuit against AVC claiming that its firmware could easily be changed to alter vote counts, the firm insisted that using better seals (physical security) would be enough to resolve the problem. The firm then forced Appel to redact his public report on their machines, claiming he was exposing "trade secrets."

It is evident that we must fix this system. And finally, there may be the political will necessary to make major changes. After the 2016 election, intelligence agencies concluded that Russia did attempt to intervene in the election and both parties have said they want to improve security in the future.[86] A number of viable proposals have been put forth, many of which we discussed in the previous chapter, on countermeasures. In selecting our ultimate recommendation, we want to focus on balancing reliability against practicality. We believe the best option for improving the reliability and integrity of U.S. elections is to replace direct-recording electronic voting machines with open source voting machines that implement voter-verifiable elections (such as Eperio).

Open Source voting will be discomforting for many — we are putting our code up for all the hackers in the world to see — but it is necessary. Especially when the stakes are this high, any attempt at security through obscurity only weakens security researchers who are trying to defend election integrity and strengthens state actors who have the resources and capabilities to exploit vulnerabilities in the system. These attackers will find and use vulnerabilities, with or without access to the code base. If the open source community can build a single voting machine that is subject to constant review by researchers, we will be much safer. This move might also provide reassurance to millions of U.S. voters.

The Eperio system (or a similar voter-verifiable design) should be used to enable voter verified elections. The Eperio software module already exists and is easily implemented. The cryptography behind the solution is relatively straightforward, which will enable at least trained engineers to understand the protocol. Any voter will be able to check that his/her vote was logged correctly (which is an important feature in any new election system) and mutually distrustful parties will be able to audit election results in a reliable way.

Some counties will choose not to buy open source voting machines. In this case, FEC should specify a much stricter testing protocol for voting machines produced by traditional manufacturers. All voting machines should be required to undergo an independent review of their hardware and source code by third-party researchers before being approved. States should withdraw their approval for voting machines with known problems and require a complete redesign, where necessary. (For an example of how this might be implemented in practice, see California's temporary ban on Hart InterCivic machines, discussed earlier).

In conclusion, we re-iterate that a wholesale revamping of the technologies used in U.S. elections is necessary to resolve our serious concerns about the integrity of U.S. elections.

Personal Research

Based on my newfound understand of voting machine vulnerabilities and attacks, I decided to conduct some personal research to 1) Verify experimentally the claims made by voting machine researchers, 2) Improve my understanding of the on-the-ground context for voting machine attacks, and 3) Prove that hacking a voting machine requires no more expertise than that which might be learned in a Bachelors Degree program in Computer Science.

Hacking a Sequoia AVC Edge

Andrew Appel bought an AVC Advantage voting machine on GovDeals.com to expose its security vulnerabilities. I decided to find

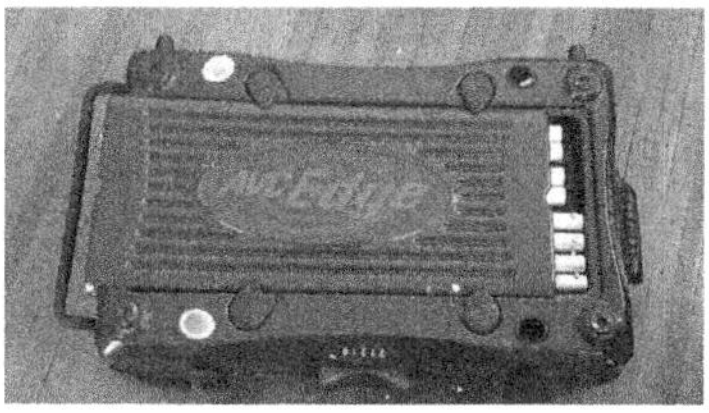

Figure 30: *AVC Edge straight out of the box*

Figure 31: *AVC Edge after setup is complete*

out if this security risk had been addressed by authorities and found that no action had been taken. Thus, I decided to buy my own voting machine to get hands-on experience with the device, learn more about how it works, improve my understanding of how voting machine attacks work, and finally attempt to execute some of these attacks "in the wild." In the past few months, more than a dozen voting machines have been sold through GovDeals.com. Unfortunately, none were for sale when I conducted my experiment. However, I found numerous voting machines were listed on eBay, and bought an AVC Edge voting machine from a government surplus dealer for $139.99, including shipping. This was fortuitous because the AVC Edge is used in Pennsylvania, where I am a registered voter. The AVC Edge is a standard direct recording electronic voting machine with a touch screen interface.

Setup

I started by opening the voting machine, which came in a large, suitcase-like box. I needed to remove four metal legs and set them up on the machine. Then I unplugged an AC Adaptor from my desktop computer and used it to power-up the AVC Edge. Finally, I fed the printer paper into the printer to get the machine totally up and running. Next, I set up a mock election, running through ballot definition and printing the "test" ballots to the VVPAT. All of this worked as expected and helped strengthen my understanding of how the voting machine works.

Figure 32: *The Control Panel for the AVC Edge*

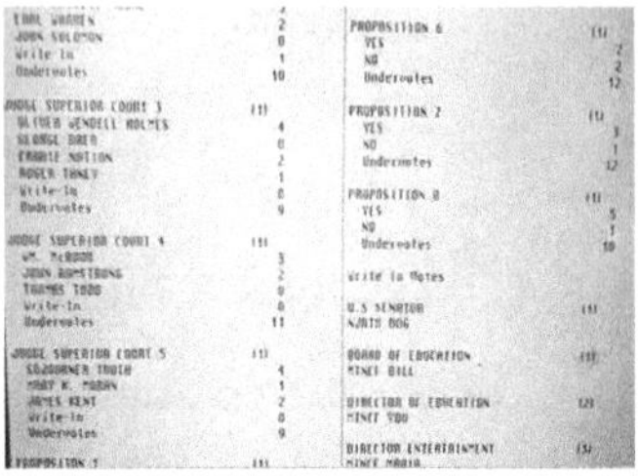

Figure 33: *Setting up AVC Edge ballot definitions*

Attacks

I attempted to attack the voting machine in three ways. First, I showed that it is trivial to set up a screen calibration attack against the AVC Edge and incorrectly calibrate the machine to deny service. Next, I unscrewed the voting machine casing and got access to the motherboard in under 10 minutes. I showed than an attacker could potentially replace any hardware / firmware (I did not make any such changes because I have no experience writing to ROM chips). Finally, I showed that any voter or poll worker could quickly access the results cartridge, for the purpose of denying service to voters, or to modify the vote count, using a homebrew PMCIA card reader/writer device. This demonstration showed that with my attack framework in mind — access control attacks, firmware attacks, and memory card attacks — I could dynamically attack any voting machine without a priori knowledge of its specific flaws.

In conclusion, my individual experience building, playing with, and ultimately attacking the AVC Edge direct-recording electronic voting machine demonstrates that it is easy to attack an arbitrary U.S. voting machine. The experience highlighted how different parts of the voting machine work — such as the access control cards, the user interface, the voter activation buttons, the VVPAT printer, the results cartridge, and the motherboard/firmware — and how to take advantage of these interrelationships to impact an election.

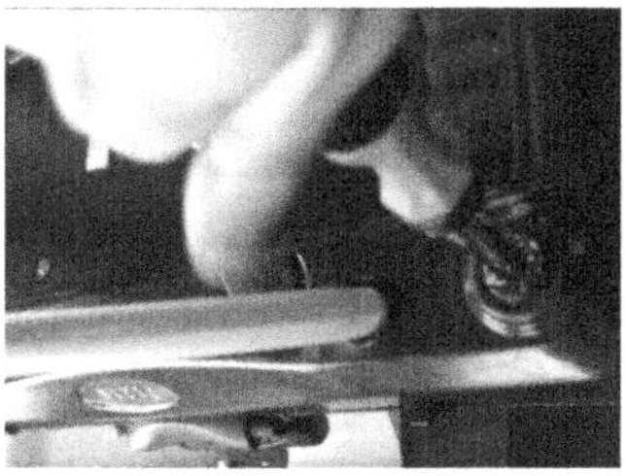

Figure 34: *Unscrewing the AVC Edge casing to gain access to the motherboard*

Figure 35: *The motherboard of my AVC Edge*

Figure 36: *Voters have easy access to the memory card with election results.*

Figure 37: *Launching a screen-calibration attack on the AVC doesn't even require the election to be set up*

You can see a short video showing how I bought, built, dissected and attacked the AVC Edge on YouTube, here: https://youtu.be/nC7ui8qdxnw. This is a clear and easy to follow application of the research in this book (albeit, amateur compared to many of the experts cited thus far).[87]

Interview with Steve [1]

Real world election volunteers and officials have unique knowledge and perspective on the election process and possible vulnerabilities. Little work has been done to integrate their perspectives into our understanding of voting machine security. Dr. Appel notes that one reason for the lack of research in this domain is that election officials

1 Steve is a pseudonym

rarely agree to cooperate with researchers, who they often view as antagonists and opponents, rather than as collaborators.[88] To bridge this gap, I set out to interview an election official so that I could learn more about the defense mechanisms currently in place to protect voting machines and evaluate the feasibility of the attacks discussed in this paper.

Finding an election official / volunteer to interview was not difficult. I had ample opportunity to discuss my research with Steve, a staff member at the University of Pennsylvania, where I did my undergraduate degree. I soon learned that Steve — who I was friends with throughout my time as an undergraduate at Penn — is involved in local Philadelphia politics as a volunteer and has served on multiple occasions as an election judge. Steve is an immigrant who has worked at Penn for almost 20 years. During this time, he took advantage of a University program that allows employees to earn degrees from Penn to complete a Bachelor's Degree in Political Science. Steve agreed to fill me in on specific details about the vetting process for election judges, the process of election set up, and his view on the attack vectors presented in this book.

1. No Vetting of Election Judges: Steve attended a political event on Penn's campus where he met the local election commissioner. After the talk, Steve signed up to be a political volunteer. A few weeks later, he was invited to a local courthouse to sign a document permitting him to be an election judge. Steve does not believe a background check was conducted before he was hired for the role. Likewise, Steve served alongside two election volunteers, who served as his "assistants" in judging the election. Based on his conversation with these volunteers, he believes they signed up because the job paid well — the county election commission paid volunteers $90 for the approximately 14-hour time commitment to monitor elections. Even if his volunteers were being vigilant at first, by the end of the long day, the volunteers' overall level of attention fell. Steve attributed this to their long hours at the polls and the monotonous nature of the task.

2. No Physical Security for Voting Machines: Steve was assigned to a precinct in which voting was conducted in the home of a local firefighter. According to Steve, the fireman stored all of the voting machines in his basement during the week before the election. The voting machines were not locked and anyone with access to the basement would have had access to the voting machines. Likewise, Steve noted that voting machines used at the University of Pennsylvania are often left unattended in the lobby of high-rises, which are student apartment buildings, in the week before an election. In this case, all Penn students would have had access to the voting machines.

3. Election Officials Don't Understand How Voting Machines Work: According to Steve, he and his colleagues received their training on how to use the voting machines at their precinct the morning of the election. Steve and his colleagues were not trained by independent election officials or by employees of AVC, the manufacturer of the voting machines. Instead, the local fireman (in whose home the voting would take place) trained the volunteers on how to use the voting machines. This lack of training would seem to confirm Dr. Appel's view that user interface attacks would be easy to carry out for a sophisticated attacker.

4. Privacy May be Compromised: I asked Steve about the sign-in book used during elections to keep track of who has voted. These books are often considered the "hard truth" in elections, and an audit can compare the number of voters registered by the voting machine to the number of voters who signed-in to the polling location. Unfortunately, these audits are rarely conducted. Steve noted two potential threats to election security created by these sign-in books. First, he noted that volunteers don't usually pay attention to which line a voter signed on, so double voting is possible if a voter 'accidentally' signs in on the wrong line. Importantly, Steve noted that in addition to the sign-in book, the polling station maintained two blue-books, which volunteers associated with each party (Steve was independent, but his two assistants were a Democrat and a Republican) used to keep track of the names of voters, in order of when they voted, as well as their political party. Steve realized that if this log were compared to the VVPAT, which is the paper trait generated by the voting machines, it would be relatively easy to figure out who voted for whom.

5. Physical Security of Memory Cards is Not a Priority: According to Steve, many election volunteers believe the VVPAT readout is the official vote count. Specifically, volunteers spend a significant amount of time counting the VVPAT readout and reporting the results to central election authorities. Steve did not appear to know that countywide election results are actually determined on the basis of the vote counts stored on memory chips within direct-recording electronic machines. Few precautions are taken to protect the physical security of these chips, which can easily be modified using a homemade device, discussed in Chapter 4.

There are many election officials who argue that attacks conducted in a lab would not be viable "in the wild" during an actual election. My interview with Steve, who is an election volunteer, indicates that security during an actual election may less stringent than we imagine. It is possible that attacks carried out in the lab do not overstate the number of attacks possible during an election, but actually *understate* the number of possible attack vectors, because they fail to account for opportunistic attacks made possible by poorly implemented security mechanisms. One easy example illustrates this point: Most threat models assume two types of attackers, insider threats (volunteers) and outsider threats (voters). In fact, in the case of Steve's precinct, the firefighter who was selected to store all of the voting machines in his house would be an ideal attacker and pose an extreme insider threat because he had end-to-end access to the technology and people involved in his precinct's election.

From: Cahn, David S [redacted]
Sent: Monday, March 06, 2017 11:55 AM
To: [redacted]
Subject: FOIA Request for Access to Voter Sign-In Book

Dear [redacted]

My name is David Cahn and I am a Senior at the University of Pennsylvania, a Philadelphia resident and registered Pennsylvania voter.

When I voted in November 2016, I signed-in at my polling location (Ward 27, Div 11). I would like to request access to this sign-in book for the purpose of academic research. I would also like to request the precise vote totals for my precinct. My intent is to compare the number of sign-ins to the total number of votes cast (a labor intensive audit that is not usually conducted).

I understand that you are the Open Information Officer for the Mayor's office. I met Mayor Kenney in December when he visited my Urban Fiscal Policy Class and he discussed the importance of transparency.

I hope you'll help me with this request. If you are not the correct person to handle this request, then please let me know who the correct person would be.

Thanks in advance.

Sincerely,
David

--

David Cahn
The Wharton School | Penn Engineering

Figure 38: *FOIA request by David Cahn to the City of Philadelphia*

Freedom of Information Act Request

We have discussed extensively already the lack of election auditing and its deleterious effect on election security. My personal experience voting in Pennsylvania (Ward 27, Div 11) on November 8, 2016 is consistent with this claim. My experience had all the hallmarks of denial of service attack. Specifically, many of the machines at my precinct were non-functional and the employees were incapable of fixing them. As a result, I waited 3 hours and 20 minutes to vote. I watched personally as dozens of voters decided not to vote, and specifically heard students say, "I need to go to class and can't wait three hours to vote," as they left.

CITY OF PHILADELPHIA

LAW DEPARTMENT
ONE PARKWAY BUILDING
1515 ARCH STREET
PHILADELPHIA, PA 19102

Sozi Pedro Tulante
City Solicitor

Jill I. Freeman
Deputy City Solicitor
(215) 683-5457
(215) 683-5069 (fax)
Jill.Freeman@phila.gov

March 13, 2017

VIA ELECTRONIC MAIL

David Cahn
The Wharton School
[redacted]

Re: Cahn CP 2017-0337i

Dear Mr. Cahn:

Thank you for writing the City of Philadelphia ("City") with your request for information. A copy of your request is attached. This letter serves as an acknowledgement of receipt of your request and notice of the City's Open Records Policy pursuant to the Right to Know Law.

Please note at the outset that written requests for records to state and local agencies in Pennsylvania are generally governed by the Commonwealth's Right to Know Law (the "Act" or "RTK Law"), 65 P.S. §§ 67.101-67.3104. In accordance with the Act, the City has promulgated an Open Records Policy, which requires the standard statewide Right to Know Form be used for the submission of written requests pursuant to the Act ("RTK Requests"). *See* 65 P.S. § 67.504(a); *Dekok v. Lancaster County*, AP 2010-1197, 2011 PA O.O.R.D. LEXIS 25 at *5 and n.1 (Pa. O.O.R. Jan. 17, 2011) (holding that in accordance with the Right to Know Law, an agency may adopt a policy requiring the use of the standard statewide form established by the Office of Open Records in order to file a Right to Know Request); *accord Pa. Gaming Control Bd. v. Office of Open Records*, 48 A.3d 503 (Pa. Cmmw. 2012). As your request was not submitted on or with the standard statewide form, the City does not consider your request to be a RTK Request pursuant to the Act and will process it as an informal request.

You do not have to resubmit your request for the City to respond unless you are specifically informed otherwise.[1] The City is continuing to process the request you have submitted as an informal request outside the Act and will be in touch with you further to respond to your request.

However, if you would like to file a RTK Request, please submit your request on or with the standard statewide Right to Know Form and address your request to the appropriate open records officer in accordance with the City's Right to Know Policy. **Please include this notice with your RTK Request.** Please note that if you resubmit your request as a written request pursuant to the Act the City will deem your informal request to have been withdrawn.

The City's Open Records Policy can be found here:

http://www.phila.gov/privacy/pdfs/FinalCityOpenRecords.pdf.

The statewide form appears at the end of the City's Right to Know Policy. If you require a hard copy of the form and/or policy, please let me know and one will be provided.

The City reserves the right to assert any and all grounds of denial in its final response to your request.

Respectfully,

Jill I. Freeman
Deputy City Solicitor

[1] The City reserves the right to require a formal RTK Request be submitted pursuant to the Act before releasing records.

Figure 39: *FOIA Response from the City of Philadelphia*

At my precinct, the vast majority of voters were University of Pennsylvania and Drexel University students. Turnout in this district was considered critical to the election outcome; Hillary Clinton came to Philadelphia the night before the election to campaign there. Former Pennsylvania Governor Ed Rendell was handing out donuts to people waiting on line to vote.[89] Clearly, established political operatives felt this was a critical precinct, and it's no wonder why — Pennsylvania is a swing state and Philadelphia leans Democratic. College students in our precinct historically voted for Democrats at a rate of 4:1, according to the *Daily Pennsylvanian.*[90] We have extensively discussed why such a precinct is an ideal target for a denial of service attack; the attacker knows that the clear majority of voters who will be affected by the attack are in the opponent's party. Likewise, denial of service attacks are difficult to detect (because they are often indistinguishable from routine hardware/software failure).

With Pennsylvania ultimately playing an outsized role in the 2016 election, the question of whether my ward — Ward 27, Division 11 — was the subject of a denial of service attack is not a trivial one. To help answer this question — and counteract the lack of an official audit, which according to Dr. Appel, are quite rare, I filed a request under the Freedom of Information Act (FOIA) to gain access to voter sign-in books for my precinct, which are a public record. My goal was to count the number of sign-ins in the book and compare that number against the number of votes tallied in the precinct, according to official reports (which are based on memory chips). This audit would uncover explicit voting machine attacks. Unfortunately, there is no way to test for denial of service attacks without explicit access to the precinct's voting machines.

At the time this book was submitted for publication, my Freedom of Information Act request is still pending review by the City of Philadelphia. The City must respond to my request eventually, but it has repeatedly delayed responding.

	Access Control Attacks	Firmware / Malware Attacks	Memory Card / Server Attacks
ES&S	PEB attacks with IR communication on Palm Pilots	Buffer overflow attacks enable voting machine virus	EVEREST Team successfully overwrites unprotected memory on M100 Optimal Scan memory cards
Diebold	Homebrew smart cards enable double-voting	Code insertion attack on Diebold machine enables attackers to take over the machine	Failed cryptography (included plain-text, hard-coded secret key) enable attackers to modify vote counts
Sequoia	User interface attacks enable disenfranchisement and double voting	Firmware overwrite attacks enable attackers to change the vote count & return oriented programing allows voting machine virus	Appel et al. design homebrew design to read and write arbitrary memory to AVC Edge memory cards
Hart InterCivic	Hardware attacks on the JBC enable DOS attacks and double voting	Firmware replacement gives attackers control of the machine	Mobile Ballot Box (MBB) uses ineffective cryptography, enabling after-the-fact changes to voting results
Other	Relay attacks on RFID-enabled smart cards		

Figure 40: *Voting Machine Attacks by Manufacturer and Category*

Access Control Attacks	Firmware / Malware Attacks	Memory Card / Server Attacks	New Voting Machine Design
Use cryptographically secure authentication	Guarantee firmware integrity	Use strong encryption to protect I/O devices	Only use optical scan voting machines
Overhaul physical security during elections	Conduct parallel testing of voting machines to check for malware	End security through obscurity in terms of memory chip encryption	Use a voter verifiable election machine (visual crypto-receipts, Eperio, or MarkPledge)
Create a simplified UI to avoid user interface attacks (no complex button sequencing)	Use software independent design so that changes to software can be detected in hardware	Always include a paper audit trail as a backup (to compare against memory chip vote count)	Use open source voting machines in all elections
Implement biometric tests for access control (not recommended)	Add digital signatures to all software elements for integrity checks	Randomly audit voter book logs to compare total number of votes against total number of voters	Test Web/SMS/online voting (not recommended)

Figure 41: *Countermeasures by Category*

Manufacturer	Year	State	Description
ES&S	2016	North Carolina	Memory cards have corrupted memory
iVotronic	2016	Texas	Candidate preferences ignored by voting machines
Sequoia	2016	Pennsylvania	Voting machines fail to record votes
ES&S	2016	Michigan	59% of voting machines misallied votes
Sequoia	2016	Utah	Improperly coded memory card leads to 75% failure of voting machines
U.S. Election Commission	2016	Washington, DC	U.S. election commission is hacked using an SQL injection
WinVOTE	2014	Virginia	57 precincts reported that their voting machines broke due to power issues
Sequoia	2008	New Jersey	Irregularities caused by user interface errors on 38 machines
Hart InterCivic	2004	New Jersey	In 21 precincts in Orange County, more ballots were cast than voters
Sequoia	2004	Florida	A number of counties report difficulties retrieving vote counts from PCMCIA cards
Diebold	2003	California	17 Diebold machines use unauthorized software, according to a security audit
iVotronic	2002	Florida	Hundreds of DREs report 0 voters

Figure 42: *Summary of Recent Voting Machine Issues*

Taxonomy of Attacks and Countermeasures

The following Taxonomy lays out a simplified overview of the attacks and countermeasures presented in this book. It offers an effective way of mentally organizing these attacks and extending these approaches to new voting machine designs and technologies. The taxonomy of countermeasures offers a mechanism for generalizing many of the voting machine defenses discussed in this book. Finally, we offer a brief summary of historical voting machine attacks, which can be sorted by state, year and manufacturer.

CHAPTER 7
WHO ARE THE HACKERS?

"It is the assessment of the intelligence community that Russia's goal here was to favor one candidate over the other, to help Trump get elected" — The New York Times

Now that we have finished discussing how to hack an election — and what the U.S. government can do to protect the integrity of our elections — we will turn to two final topics, of substantial importance. First, who is trying to hack U.S. elections and how credible of a threat do they pose to our safety? Throughout this book, we have discussed a hypothetical attacker, and often referred to state-actors as being capable enough and rich enough to invest the resources and time needed to carry out a full-scale attack. Here, we will investigate those claims. In Chapter 8, we will turn specifically to controversy surrounding the 2016 election and discuss its implications for election security in 2018 (and what we can do about it).

In this chapter, we expand our "threat model" to discuss the specific actors we must defend against. Threat modeling is used in cyber security to understand the parameters of the attack being defended against and helps us frame and understand the attacks to follow. The primary element in our threat model is understanding who we are defending against, and what their goals are. If we understand our attackers, then we will be able to think like them, anticipate their attacks, and build appropriate defenses.

Chinese and Russian attackers pose the most serious threats to U.S. election security. State-actors are well-funded, technologically sophisticated, and have ample motivation to manipulate U.S. elections to achieve their foreign policy goals. When it comes to election security, we need to be concerned with five groups of attackers: Russian government-supported hackers, Chinese military hackers, rogue actors like North Korea and Iran, black hat hackers selling their services for profit, and amateur hackers, who wreak havoc on elections for fun.

The Russian Threat

By far, the most sophisticated state actor involved in election hacking is the Russian government and its cronies. Russian hackers have been linked to some of the most devastating corporate cyber-attacks in U.S. history, and led a campaign to discredit Hillary Clinton — and support Donald Trump — during the 2016 election.

To understand the Russian cyber threat, we must go back to 2007. In an address before the Annual Security Conference in Munich, Putin laid out a doctrine that would govern the next decade's military buildup. Putin claimed that the West was engaging in illegal cyber warfare and declared that he would build his own information warfare capabilities to combat NATO. At the time, few paid heed. Today, we recognize that this was a warning shot fired across the Atlantic, and that the escalation Putin promised in 2007 is the same strategy that finally culminated in a full-scale cyber campaign against the United States in 2016.[91]

"Vladimir Putin threw down the gauntlet to the west in a confrontational speech on Saturday, attacking what he called "illegal" US unilateral military action and arguing it had made the world more dangerous," reported the Financial Times, in February 2007. "To an audience that included Robert Gates, US Defense secretary, John McCain, US presidential contender, and a group of Washington lawmakers, Mr. Putin declared the end of the unipolar world, which he described as a failure for the world and the US itself." [92]

"We don't have enough force to resolve anything comprehensively," Putin is reported to have said. "We are forced to think about guarantees of our security."

In 2008, Russia formalized its foreign policy goal in "The Foreign Policy Concept of the Russian Federation," noting that the world had become more digitally connected, and the "well-being of nation states would become dependent on data and its rapid movement."[93] At the time, Russia was estimated to have 7,300 cyber warfare experts employed by the army, with a cyber warfare arsenal including: a large advanced botnet for DDoS and espionage, electromagnetic pulse weapons (non-nuclear), compromised counterfeit computer software, advanced dynamic exploitation capabilities, wireless data communications jammers, cyber logic bombs, computer viruses and worms, cyber data collection exploits, computer and networks reconnaissance tools, and embedded Trojan time bombs (suspected).[94]

With a new arsenal of weapons in development, Russia found plenty of opportunities to test and perfect its capabilities in regional conflicts.[95]

In April 2007, Estonia became the first target. The conflict arose out of seemingly innocuous circumstances. Estonians removed a Soviet war monument from the center of the capital to a military cemetery. The decision led to massive riots and looting, specifically among the ethnic Russian population. Russians admired the monument, believing it represented their wartime sacrifices, whereas the Estonians opposed the monument, seeing it as a symbol of Russian occupation. The conflict soon turned violent, with Russian youth groups blockading the Estonian embassy, and the Russian government calling for Estonian leaders to step down.

"For a small, high-tech country such as Estonia, the internet is vital. But for the past two weeks Estonia's state websites (and some private ones) have been hit by "denial of service" attacks, in which a target site is bombarded with so many bogus requests for information that it crashes," reported *The Economist*. Within weeks, a deal was brokered by Germany, and the conflict formally ended. Not so for the cyber war. In the weeks and months following the deal with Russia, Estonian websites continued to be vandalized by the Russians, and attacked with Denial of Service attacks.

These attacks did not go unnoticed in the West. But lacking any precedent for such acts of war, NATO leaders admitted that they did not know how to respond. "If a member state's communications center is attacked with a missile, you call it an act of war. So what do you call it if the same installation is disabled with a cyber-attack?" asked a senior Brussels official. At the time, Estonia attributed the attacks to two groups: initially, many attacks could be directly traced to Russian government computers. However, over time, the attacks migrated to hundreds of computers around the world. We now know these networks to be "botnets," anonymous troves of computers whose owners run scripts posted to Russian-language sites, linking them to a network of attackers. Each attacker invests negligible computing power, but collectively, the botnets can cause real damage, such as disabling target websites. Botnets have become increasingly sophisticated; many botnet participants claim that their computers were hijacked by viruses that linked them to the botnet without their knowledge.

Today, an entire criminal industry has sprouted around these botnets. Many researchers believe that botnets are the most profitable element of the cybercrime industry, with "hackers for hire" infecting thousands of computers with malware, linking them to a custom botnet, and then selling botnet time, per hour, to the highest bidder. In Estonia's case, this means that the attacks are not traceable to the Russian government, even as Russian money funds botnet attacks.

As the attacks escalated, Estonia had no choice but to disconnect its computers from the foreign Internet. *The Economist* noted that the defense mechanism would cause even more damage to the Estonian economy that Russian sanctions. "We are back to the stone age, telling the world what is going on with phone and fax," explained an Estonian Internet expert. Mikko Hyppönen who monitored the attacks, said that the best defense to botnets of this sort would have been strong networks of servers, distributed globally, to limit vulnerability.[96]

If Estonia gave Russia the opportunity to beta-test its new cyber warfare capabilities, Georgia was the first real demonstration of the potential damage cyber-attacks could cause. The Georgian conflict, which involved full-fledged military escalation, involved a dispute over the Georgian province of South Ossetia. Russia claimed the providence was part of Russia. In the weeks that followed, a full-fledged Russian cyber war ensued. The Georgian central government website was attacked in a denial of service attack, the Ministry of Foreign Affairs, Ministry of Defense and large commercial websites were attacked as well.

"A cyber warfare campaign by Russia is seriously disrupting many Georgian websites, including that of the Ministry of Foreign Affairs," the Georgian Ministry of Foreign Affairs said in an official statement. Cyber security analysts said the Russian Business Network (RBN), a group of hackers with ties to the Russian government, were responsible for the Georgian attacks. Jart Armin, who tracks RBN, said the attacks were routed through servers that "are well known to be under the control of RBN and influenced by the Russian Government." The Shadowserver Foundation, a watchdog, confirmed that "We are now seeing new attacks against .ge sites - www.parliament.ge and president.gov.ge are currently being hit with http floods." Just as the Russians had practiced in Estonia, a full-scale distributed denial of service attack was in progress. The conflict escalated when Deutche Telekom stepped in to reroute Georgian traffic through its servers, resulting in a tug-of-war between the German telecommunications giant and the Russian, state-sponsored hackers. [97]

Following its successful campaigns in Estonia and Georgia, Russia scaled up its cyber warfare tactics, using them in global geopolitical conflicts, such as the wars in Ukraine and Syria. "It comes at different times, and in different forms," reported Foreign Policy Magazine in

2015. "But as they have charted the war in southeast Ukraine over the past year, drones flown by the Organization for Security and Co-operation in Europe have run into the same problem: Russian troops on the ground are jamming them into virtual blindness."

Lt. Gen. Ben Hodges, commander of U.S. Army units in Europe, said Russian electronic-warfare capabilities were "eye watering." Beginning with the Crimean conflict in 2014, the U.S. government had a rude awakening; they were not prepared to combat the Russian cyber threat. "You can't but come to the conclusion that we're not making progress at the pace the threat demands," Ronald Pontius, deputy to Army Cyber Command's chief explained. In Crimea and Ukraine, the U.S. found that military radios and phones were inoperable, and that drones were being jammed with military-grade technology.

"They have companies, they have battalions, they have brigades that are dedicated to the electronic warfare mission," Col. Jeffrey Church, the Army's chief of electronic warfare, told Foreign Policy Magazine. Those units are deploying "with specific electronic warfare equipment, with specific electronic warfare chains of command." By contrast, the U.S. had just 813 soldiers for the Army's electronic warfare team at the time, with only 1,000 positions authorized. The soldiers are typically distributed two per battalion. A battalion consists of 300 to 800 total soldiers. Given the low ratio of electronic warfare specialists in U.S. battalions, Church wonders "how do you maintain in a high-intensity environment against a peer enemy?... "There's too much to do for those guys in a battalion." [98]

The U.S. has not answered this question. In an assessment, the U.S. Army's Foreign Military Studies Office said that Russia "does indeed possess a growing EW capability, and the political and military leadership understand the importance" of such warfare. "Their growing ability to blind or disrupt digital communications might help level the playing field when fighting against a superior conventional foe."

In recent years, Russian cyber warfare capabilities have matured considerably. A World Bank study found that Russian has more than 1 million software specialists involved in research and development. There are 40 large cybercriminal rings operating in Russia. Crucially, Russia has provided these criminals with immunity, allowing the cyber-crime industry to grow considerably.[99] Infosecurity Magazine estimated that Russian cybercriminal groups had recruited an additional 1000 engineers between 2012 and 2015. These gangs are funded by scams they run on the Internet, and by selling their services as "hackers for hire."[100]

Russian attacks have expanded from ad hoc attacks on its neighbors to a "sustained campaign of cyber-attacks targeting democracy and critical infrastructure in the West," according to British Defense Secretary Michael Fallon. This "sustained campaign," included a 2015 attack against the German parliament, the Bundestag. The Russian attack infected 20,000 German computers, used by government employees and Bundestag members/staff. The German Federal Office for Computer Security (BSI) said it would have to replace all the effected computers, which would cost millions of euros. "This is the biggest cyber-attack on the Bund, on the German parliament," said Bernhard Kaster, secretary of the Christian Democratic Union and Christian Social Union political bloc. [101]

Bulgaria too, found itself to be a repeated target of Russian aggression. The country was attacked during a 2015 referendum and local elections, which the Bulgarians attributed to Russia. Rosen Plevneliev, Bulgaria's president at the time, warned that Russian influence was growing in Bulgaria and all throughout Europe, telling BBC that the continent needs to take a stronger line against the Russians. "The game in Europe today is not to have a full-scale war and to shoot against your enemy, but the game of Mr. Putin is to make other countries dependent," Plevneliev said. He described Russia as "dangerous and unpredictable" and described Russia's repeated aggression and confrontation as "cold peacetime."

According to Plevneliev, Russia wields several geopolitical tactics, which it uses to spread its influence across Europe, including: funding populist political parties and movements, extensive propaganda and cyber-attacks to destabilize opponents. He described the 2015 Russian cyber-attack as the "most heaviest and intense cyber attack that has been conducted in south-east Europe." The distributed denial of service attack, like the ones used in Estonia and Georgia, targeted the election commission, presidency and government institutions. Unlike in Georgia, the attack was launched on the day of a referendum and local election — a blatant attempt to subvert Bulgarian democracy.

"The same organization that has attacked the Bundestag- stealing all the emails of German members of parliament - the same institution that has attacked NATO headquarters, and that is the same even that has tried to influence American elections lately and so in a very high probability you could point east from us," Plevneliev said of Russians persistent attacks and meddling with global institutions and elections. Western security analysts believe all of these attacks were attributable to the "Fancy Bears" or APT 28, a group of hackers linked to the Russian government, BBC reported. [102]

It is with this extensive context that we turn to Russia's next attack — and the attack that will have consequences for our story. That attack is Russia's campaign to influence the outcome of the 2016 election, beginning with its theft of more than 19,000 emails from Democratic Party officials. "We believe, based on the scope and sensitivity of these efforts, that only Russia's senior-most officials could have authorized these activities," said the Office of the Director of National Intelligence and the Department of Homeland Security (DHS) in a joint statement.

A CIA report found that Russian operatives did work to covertly influence the outcome of the 2016 election. In early 2017, the *New York Times* reported that intelligence officials had "high confidence" that Russia was involved in hacking related to the elections. "It is the assessment of the intelligence community that Russia's goal here was to favor one candidate over the other, to help Trump get elected," a source told the *New York Times*. "This is something that we have to make sure that we're on guard for," CIA Director Brennan said on Face the Nation. "Not just for our national security purposes but also for making sure that our system of government here is going to be preserved."

Before we move on, it is important to note that none of the evidence indicates that Russian interference materially impacted the outcome of the U.S. election. Certainly, the extent of Russian cyber-attacks and their impact is still being unraveled. We'll come back to this topic at the end of this book, as we consider implications for the upcoming 2018-midterm elections.[103]

Putting a Face to the Name

If Russia has succeeding in building a massive corps of hackers to do its bidding, the next obvious question in building our threat model should be — who are the attackers? We can only learn so much from nicknames like the "Fancy Bears" or Advanced Persistent Threat (APT) 28. We need to put a face to the name, to truly understand the psychology and motivations of the attackers. We know that Russia has plenty of technical talent coming out of its universities, which are regarded as some of the best engineering colleges in the world. But who are the hackers, and how are they recruited?

In December 2016, a groundbreaking *New York Times* interview gave us some insight into what the profile of a Russian hacker might look like, and how Russia attracts these talented hackers to do its bidding. The *Times* interviewed Aleksandr B. Vyarya, "bearded, bespectacled computer programmer who thwarted hackers." Vyarya said he was approached by the Russian government, and asked to join a new cyber warfare taskforce. Vyarya said that the government was launching a "sweeping overhaul" of the military, to make cyber warfare a central tenet of Russia's foreign policy. He was to a join a new unit.

Vyarya was an attractive candidate for many reasons, but chiefly because he understood how distributed denial of service attacks work.

In the private sector, he helped defend companies against these attacks. His clients included a newspaper called Vedomosti; a television station called TV Rain, and the website of opposition leader Aleksei Navalny, the opposition leader. Varya told the Times he was invited to join Vasily Brovko, of the military contractor Rostec, to Bulgaria for the demonstration of a new denial of service technology. Brovko quickly invited Vyarya to lead a new denial of service initiative, saying that his company would buy the Bulgarian software.

"This is against my principles — and illegal," Vyarya told the Times. After declining the offer, Vyarya was surveilled, and an acquaintance in law enforcement advised him to flee the country. "As soon as we saw what was on the table, [Vyarya] was given direct instructions to return to his hotel and stop all contacts," his former boss, Aleksandr Lyamin of Moscow cyber defense company Qrator told the Times. Brovko and Rostec denied the account, called Vyarya "mentally ill" in a statement.

Fearing retribution by the Russian government for his refusal to sign-up with the government contractor, Vyarya fled to nearby Finland. Vyarya was just one of many of a need breed of cyber warriors, sought after by the Russian government. By offering hackers jobs and perks, as well as an implicit threat of retribution if the hackers decline their offers, the Russian government is working to co-opt programming talent in the country to help it achieve their foreign policy goals. Analysts believe the new escalation is attributable to the Gerasimov Doctrine, proposed by General Valery Gerasimov. Gerasimov, who was appointed Russia's Defense Minister in 2012, argued that there is no longer a clear distinction between war and peace. As a result, "covert tactics, such as working through proxies or otherwise in the shadows, would rise in importance." [104]

The Russians have cast a wide net in their search for new programmers, seeking to recruiting programmers from universities, the private sector, and the extensive Russian underworld, where hackers for hire thrive. Russia is one of the only countries in the world that offers relative immunity to cybercriminals.[105] When Dmitry Artimovich was arrested for a mass spamming campaign, his cellmate explained that the Federal Security Service (F.S.B.) would allow him to avoid a trial, if he agreed to work for them.

"It was an offer to cooperate," Artimovich told the *Times*. "Why else would you work for the government?...The salaries are tiny. But if you do something illegal, and go to prison for eight or nine years, the F.S.B. can help you." (Artimovich chose to go to trial and served a year in a penal colony). [106]

"For more than three years, rather than rely on military officers working out of isolated bunkers, Russian government recruiters have scouted a wide range of programmers, placing prominent ads on social media sites, offering jobs to college students and professional coders, and even speaking openly about looking in Russia's criminal underworld for potential talent," the *Times* reported.

Russia has hardly kept its new tactics a secret. The Defense Ministry has purchased video ads on Vkontakte, a Russian social networking site. In one video, a man with a rifle types into a computer — revealing the new breed of Russian soldiers. "If you graduated from college, if you are a technical specialist, if you are ready to use your knowledge, we give you an opportunity," declares the narrator. "Members of the science squadrons live in comfortable accommodation." The video jumps to a newly apartment furnished, with a washing machine.

Meanwhile, new university graduates, many of whom are subject to mandatory conscription, see the "science squadrons" as a way to avoid a more difficult army assignment. On their draft questionnaires, new recruits are asked about their knowledge of programming languages. In an expose, Russian news site Meduza became the first news agency to report on expanded recruitment efforts, highlighting posts on job websites looking for programmers with "patches, vulnerabilities and exploits." [107]

As early as 2013, the Russian General Oleg Ostapenko is said to have considered including hackers with criminal histories in its science squads. In an article entitled, "Enlisted Hacker" for the Rossiiskaya Gazeta, the government newspaper, Ostapenko was quoted as saying "From the point of view of using scientific potential, this is a matter for discussion." U.S. cybersecurity experts believe this strategy was implemented. "There have been cases where cybercriminals are arrested but never ended up in prison," said Dmitri Alperovitch, the co-founder and chief technology officer of CrowdStrike, a leading U.S. cyber security company.

What's Next?

Just how extensive are Russia's cyber warfare capabilities? Research indicates that Russian attacks during the 2016 election barely scratch the surface of Russia's offensive capabilities. In 2015, the Russian government attacked the Ukranian power grid, cutting off power for 250,000 people. [108] The attack indicates that Russia's offensive capabilities may extend beyond mis-information campaigns, email theft, and denial of service attacks. A White Paper on the Ukrainian attack concludes that Russia's offensive capabilities could cause real-world, physical damage to critical infrastructure. [109] A report by Llyod's of London concluded that an attack on the northeast power grid in the U.S. could cause a multi-week blackout, with economic losses caused ranging from $250Bn to $1Tn dollars. Economist losses are attributable to interruption of business, reduced sales, breakdown of logistics systems, and even looting/rioting, in the worst-case scenario. Attacking air traffic control systems could down commercial flights. The Mirai botnet, which is famous for taking over thousands of computers and Internet of Things devices, is thought to be capable of taking down core Internet infrastructure. [110]

The malware used against Ukraine has now been dubbed Crash-Override. In the months following the attack, researchers confirmed that with a few modifications, it could be used against the United States. "It's the culmination of over a decade of theory and attack scenarios," Sergio Caltagirone, director of threat intelligence for Drago, told the *Washington Post.* "It's a game changer." The group that built CrashOverride is thought to be involved with numerous other infrastructure attacks, including one that targeted U.S. industrial control systems in 2014.

"U.S. utilities have been enhancing their cyber security, but attacker tools like this one pose a very real risk to reliable operation of power systems," said Michael Assante former chief security officer of the North American Electric Reliability Corporation, explained. We will not focus on these doomsday scenarios in our book. Instead, we'll focus on much simpler attacks on voting machines which can be carried out with little to no training. The focus of our book is on election security. But understanding the outer limit of Russian capabilities — and the extensive damage Russian hackers could wreak on the U.S. — is critical to understanding who we're up against. [111]

It's Not Always the Russians

As we shall see shortly, the Russians are only one of many threat actors on the geopolitical stage. In fact, China, North Korea, and for-profit "hackers-for-hire" pose a significant threat to U.S. national security and election integrity. Unlike the Russians, you rarely read about these attackers (even China), in the press. In her controversial paper, "Blame It on the Russians: Tracking the Portrayal of Russian Hackers during Cyber Conflict Incidents," University of Hull Professor Athina Karatgozianni argues that Russia is often blamed for cyber attacks that it is not actually responsible for. Her point of view deserves our attention.

Karatgozianni's reviews more than 150 newspaper articles discussing Russian cybercrime. She argues that the strength of Russian hackers is often exaggerated by the media, and that the narrative surrounding Russian hackers is result of a post-Cold War dynamic in the West, where Russia is still seen as a strong and threatening opponent. Even while Russia is responsible for 12% of password theft (vs. 63% for China), there is a special "mythology surrounding the abilities of Russian hackers, where they emerge as super hackers with astonishing accomplishments."[112]

We fundamentally disagree with Karatgozianni's argument. As we shall see in the remainder of this chapter, while black-hat hackers and sophisticated state-actors like China do pose a threat to U.S. national security, Russia is the most substantial threat to U.S. election security, because of its top-down government policy of meddling in the foreign affairs of its neighbors and of Western democracies. Most non-Russian hackers are motivated by economic payoff and are therefore focused on intellectual property theft and corporate espionage. Russia may be alone in the extensiveness of the criminal network it has set up to achieve its geopolitical goals, independent of any economic payoff. In the remainder of this chapter, we will discuss the capabilities and motivations of these other hackers — China, North Korea, Iran, black hat hackers and amateur hackers. These groups will help complete our threat model.

Chinese Military Hackers

The scale of Chinese cyber warfare is extensive. "The Chinese have penetrated every major corporation of any consequence in the United States and taken information," according to former NSA Director Mike McConnell. "We've never, ever not found Chinese malware."[113]

For Chinese hackers, the incentive is simple. Intellectual property can be valuable to local companies looking to undercut their U.S. competitors, and these companies are said to pay a premium for stolen data. The impact of such attacks can be devastating, as we've seen recently in the cases of Home Depot, SONY, Anthem and JP Morgan. Nationally, their impact adds up: the Center for Strategic and International Studies estimates that intellectual property theft by Chinese hackers costs the U.S. $100 billion dollars and 508,000 jobs annually.

If hackers have developed massive expertise in the private sector, how much damage can they do in the public sector? In 2015, Chinese hackers infiltrated the U.S. Office of Personnel Management and stole millions of records. After the attacker, President Obama announced, "We are preparing a number of measures that will indicate to the Chinese that this is not just a matter of us being mildly upset but is something that will put significant strains on the bilateral relationship if not resolved." Since then, with President Trump in office, tensions have escalated. Trump has threatened trade wars with China, escalating the rhetoric as Chinese hackers increase the frequency and potency of their attacks.

Who are China's advanced hackers? A report released by Mandiant, a computer security firm based in the U.S., indicated that China's most advanced hackers, known as "Comment Crew" or "Shanghai Group" may be based out of Unit 61398 of the People's Liberation Army.[114]

McConnell, who served under President H.W. Bush and is now a senior consultant at Booz Allen Hamilton, estimates that the Chinese government alone employs 100,000 hackers — around the same number of total spies working for the U.S. government. These hackers commit economic and political attacks, which are a critical arrow in China's foreign policy toolkit. In 2014, the U.S. indicted five Chinese officers for computer hacking and economic espionage.[115]

The Communist Party is said to view IP theft as an economic policy — a way of boosting its GDP growth during times of low growth. "Indeed, as the Chinese economy enters what appears to be a protracted slowdown—if not a permanently lower rate of growth altogether—we can expect commercial cyber-attacks to increase," explained Harry Krejsa, an Asia-Pacific Analyst at the Center for a New American Security. Likewise, hacking elections is a pragmatic foreign policy tool, intended to wreak havoc on U.S. election outcomes. Hacking of voting machines can materially benefit the Chinese government, by helping elect politicians with favorable foreign policy perspectives, or by weakening U.S. democracy, which many in China see as an alternative to China's one-party rule.

While there is ample evidence that Chinese hackers have attacked U.S. companies and defense infrastructure, evidence of attacks against U.S. elections is much more limited. Since 2015, the overall number of attacks originating in China has dropped dramatically, after President Obama and China' President Xi Jinping agreed to not to conduct cyber-enabled theft of business secrets, though they did not limit hacking for political purposes. In June 2016, FireEye reported a decrease in cyber espionage from 72 China-based groups, which it attributed to factors including President Xi's military and political initiatives, the widespread exposure of Chinese cyber operations, and mounting pressure from the U.S. Government."

In a separate report, FireEye noted the importance of proactively defending U.S. elections, well before state-actors begin to launch their attacks. "The threat actor activity which facilitates these operations often takes place well before Election Day, and as we saw in the US, targets can be very diverse. The unfortunate reality today is this is a threat with which all democracies must contend," they wrote. Nonetheless, evidence from their initial testing indicates that the Chinese government may be more focused on manipulating election outcomes in neighboring states, like India, as opposed to hacking U.S. elections.[116]

Despite limited evidence that China has intervened in U.S. elections, President Trump has lambasted the Chinese extensively, most often for the role they played in hacking the Office of Personal Management, an attack that involved stealing millions of sensitive security-clearance documents. "Twenty-two million accounts were hacked in this country by China," Trump said in a news conference after the 2016 election. Around the same time, Senator Jeff Sessions declared: "The problem may turn out to be, as in the Chinese hacking of hundreds of thousands of — maybe millions — of records, it has to be handled at a political level."

"People who live in glass houses need to think about throwing rocks, because this was an act of espionage," then-Director of National

Intelligence James Clapper told the Senate Intelligence Committee.

In conclusion, while there is ample evidence that the Chinese government and Chinese hackers are involved in economic espionage against the United States, there is no clear evidence that China has hacked or plans to hack U.S. elections to achieve foreign policy goals. That said, we have two caveats. First, hacking U.S. voting machines is trivial, and so Chinese attacks may simply go undetected vs. large-scale corporate espionage, which is easier to detect after the fact. Second, the fact that the Chinese have not thus far turned their cyber warfare capabilities toward the goal of manipulating U.S. elections offers no guarantee they will not choose to do so in the future. If we leave the door open to Chinese hackers, and make it easy to manipulate U.S. elections, we must take responsibility when foreign governments eventually take advantage of our vulnerabilities.

Rogue Actors: North Korea and Iran

The long-story short on North Korea and Iran is that while they have highly advanced cyber warfare capabilities, neither appears to be focused on U.S. elections (yet). Instead, North Korea is focused on prodding its neighbor and sworn enemy, South Korea, and Iran is focused on attacking U.S. economic institutions and Israel. Both countries, despite being rogue nations with relatively little economic power, have developed world-class cyber security programs. If these countries were to follows Russia's lead — which is plausible, given their apparent goal of wreaking as much havoc as possible on the United States— it is likely that they would both pose credible threats to U.S. election security. If a college student can hack a voting machine, then both countries, with their sophisticated military hacking units and corps of programmers, could attack U.S. election infrastructure.

In the media, reports about North Korea often focus on the country's unpredictable despot, Kim Jong-un and his relationship with President Donald Trump. The two leaders have called each other "rocket man" and "dotard," respectively. Underneath the simmering tensions over North Korea's nuclear program, is a second threat that has the potential to wreak havoc on U.S. institutions: North Korea's 6,000-man strong cyber warfare unit. U.S. security leaders have described North Korea's hackers as "undeniably persistent, and undeniably improving."

In October of 2017, the *New York Times* reported that North Korea's cyber program, once considered an ill-conceived attempt by the backward nation to rival western hacking, is now a force to be reckoned with. "The World Once Laughed at North Korean Cyberpower. No More," read the *Times* headline.[117] In 2013, an attack on South Korean banks took down 50,000 computers. In 2016, North Korea came close to stealing $1Bn from the Federal Reserve (it got away with only $81MM) and took down thousands of computers, including the National Health Service in Brittan, in a well-publicized attack.

"Cyber is a tailor-made instrument of power for them," Chris Inglis, a former deputy director of the National Security Agency, told the *New York Times*. "There's a low cost of entry, it's largely asymmetrical, there's some degree of anonymity and stealth in its use. It can hold large swaths of nation state infrastructure and private-sector infrastructure at risk. It's a source of income." Speaking at the Cambridge Cyber Summit this month, Inglis added: "You could argue that they have one of the most successful cyber-programs on the planet, not because it's technically sophisticated, but because it has achieved all of their aims at very low cost."

North Korea's commercial attacks have been widely publicized. In 2014, North Korea attacked SONY, to prevent it from launching *The Interview,* a movie that mocked the North Korea dictator. The attackers released thousands of documents, including emails and unreleased films that were stolen and published to the Internet. As a result, *The Interview* was not released in theaters, and instead was offered online for download. The attack has been linked to Unit 180, a hacking unit of North Korea's military. Around the same time, North Korea hacked into a British television network to prevent a film about a nuclear scientist kidnapped in Pyongyang from being released. These attacks helped launch North Korea as a force to be reckoned with. It also highlighted just how vulnerable large corporations are to cyber-attacks.

"We learned that you really have no way of protecting yourself in any meaningful way," explained Michael Lynton, who was Sony's CEO at the time of the attack.

"Because they are such a mix of the weird and absurd and medieval and highly sophisticated, people didn't take it seriously," explained Robert Hannigan, the former director of Britain's Government Communications Headquarters. "How can such an isolated, backward country have this capability? Well, how can such an isolated backward country have this nuclear ability?"

The roots of North Korea's cyber warfare program were planted in the early 2000s, when leadership in Pyongyang recognized the potential to weaponize the Internet, and began sending promising students to receive special training, often at Chinese universities. In the late 1990s, the FBI noticed that North Koreans working at the UN had begun attending computer science classes in New York City. "If warfare was about bullets and oil until now," Kim Jong-il is reported to have told his commanders soon after invasion of Iraq War, "Warfare in the 21st century is about information."

Kim Heung-kwang, a defector who said that he was responsible for training some of North Korea's earlier hackers, explained how cyber warfare first came to North Korea via China. In an interview, he explained that a group of North Korean computer experts return from a trip to China with "a very strange new idea…Use the Internet to steal secrets and attack the government's enemies." "The Chinese are already doing it," Heung-kwang remembers being told.[118] By 1994, the Workers' Party, which rules North Korea, agreed to send 15 North Korean students to a military academy in Beijing, where they could learn how to hack. This group would form the core of North Korea's newly formed

External Information Intelligence Office. The group started by hacking websites, penetrating fire walls, and stealing data. This group launched most of their attacks from China and Japan, a trend that become formalized and permanent. Even today, most North Korean hackers don't reside in North Korea. There is much better infrastructure, in the form of web connections and IT infrastructure, abroad.

Today, India, Malaysia, New Zealand, Nepal, Kenya, Mozambique and Indonesia are all sources of North Korean cyber attacks, according to Recorded Future, a cyber security company. Sometimes, the attackers merely route their attacks through a foreign country, but often, researchers say, the attackers are now physically located in the foreign county. Researchers believe that India is the source of a more than 20% of North Korean cyber attacks. As with many teams, the initial decisions of the first Chinese cyber warriors began a trend that continues to be relevant today. Thus, in trying to understand North Korea's cyber warfare capabilities, there is a lot we can learn from the program's early days and development.

By 1996, Kim Heung-kwang says, the Korean military began training hackers "in earnest." In 1998, the infamous Bureau 121, which is now the country's most sophisticated cyber warfare unit, was formed. Like its predecessors, Bureau 121 sent new inductees to China or Russia for training. Jang Sae-yul, an army lieutenant who wrote software for war game simulations, told the *Times* that these hackers were "envied" because they had the unique opportunity to travel abroad. "They used to come back with exotic foreign clothes and expensive electronics like rice cookers and cameras," Jang said. According to Jang, when the hackers returned to North Korea, they were assigned a region or country, which they would learn to target. The most popular regions were the U.S., South Korea, and China.

"They spend those two years not attacking, but just learning about their target country's Internet," Jang said. As the pipeline for hackers became more formalized, the government began identifying hackers as early as high school, sending them to the country's top universities, including Mirim University, a military academy that specializes in computer-based warfare. Jang himself attended Mirim as an officer.

For Kim Jong Il, the early days of cyber warfare involved disrupting GPS systems. According to *Bloomberg*, Il hired hackers to develop the technology necessary to knock out foreign computers, using an electro-magnetic pulse (EMP). It was Il who is said to have first established Unit 121, the country's first cyber warrior unit. In 2004, the unit was said to have hacked the South Korean army's wireless communication system. It drew attention for testing malware and stealing money to fund future operations. In 2011, five hackers South Korean police arrested five people for allegedly working with North Korean hackers to steal millions of dollars through online games.[119]

North Korea's most famous attack came in May 2016. The WannaCry attack, launched in Britain, took down the computer systems of major hospitals, which caused ambulances to be re-routed and surgeries to be postponed. The ransomware attack was linked to a North Korea-affiliated group called Lazarus. Overall, the attack affected 300,000 computers. For each computer, Lazarus threatened to erase all the data unless they were paid $300 in bitcoin, within 72 hours. Bitcoin is the currency of choice for North Korean hackers, because it can circumvent trade restrictions and United Nations sanctions. The attack was contained only due to sheer luck — a young British computer scientist found a kill switch in the software, which he activated by purchasing a domain name on the Internet for $10.69.

Since WannaCry, North Korean attacks have accelerated. In October 2016, South Korea disclosed that highly classified military plans were stolen. These plans discussed how the U.S. and South Korea would act in the case of an armed conflict, including a plan to "decapitate" the North Korean regime by killing Kim Jong Un. Rhee Cheol-hee, a member of South Korea's ruling party, reported that 235 gigabytes of military documents were stolen from military data centers. At the time,

80% of the stolen documents had not been identified.[120] In the same month, North Korean hackers were linked to a $60MM theft from Taiwan's Far Eastern International Bank. Far from being set back by U.S. deterrence, North Korea has only gotten bolder.

In 2017, North Korea's primary target has been bitcoin, culminating in an attack on South Korea's largest bitcoin exchange, Youbit. After 8 months of persistent attacks, North Korean hackers were able to take down the exchange. Luke McNamara, a cyber security expert, told *The Daily Express* that "We have noticed North Korean hackers actually pivot to start going after places like cryptocurrency exchanges in South Korea…. The fact that it has appreciated so much in value this year means that it's attracting a lot of interest from various cybercriminals." Analysts believe that cryptocurrency attacks, like North Korea's bank heists, are being used to fund the country's nuclear program.[121]

The NSA has made significant headway in fighting back against North Korean hackers. As early as 2010, NSA hackers had penetrated North Korean cyber defenses. The NSA hackers identified web connections in Malaysia linked to North Korean hackers and penetrated Chinese networks. Using these points of entry and teaming up with South Korean intelligence officials, NSA hackers were able to get into North Korea's networks. [122] According to a recently disclosed NSA report, the effort expanded into a full-scale attempt to insert malware onto North Korean computers and gather intelligence on the Korean hacking effort, namely, the Reconnaissance General Bureau and Bureau 121, the North's secretive intelligence and hacking units.[123]

This effort created an "early warning radar" system and helped convince President Obama that Kim Jong-un's government was responsible for the SONY attack, according to the *New York Times*. Intelligence agency officials spun this decision as a major victory for their intelligence operation — Obama's decision to accuse North Korea in the SONY attack, which was the "largest destructive [cyber] attack against an American target — and to promise retaliation, which has begun in the form of new economic sanctions — was highly unusual." According to the *Times,* it was the first time the United States ever charged a government with cyberwarfare.

"Attributing where attacks come from is incredibly difficult and slow," said James A. Lewis, a cyber warfare expert at the Center for Strategic and International Studies in Washington. "The speed and certainty with which the United States made its determinations about North Korea told you that something was different here — that they had some kind of inside view." This inside view was provided by a series of implanted "beacons," which cost billions of dollars and provide surveillance, and sometimes even destructive malware capabilities.

Even as the U.S. intelligence agencies bragged about their "beacons," they begged the question: Why didn't we see this coming and stop the attack before it was launched? After all, the United States has the largest and best funded military in the world. North Korea is an isolated dictatorship with no economy to speak of and a tiny military budget. How is it that they have better hackers than us, or that they continue to win tactical victories, wreaking havoc on U.S. corporations, taking U.S. targets by surprise, and stealing millions of dollars.

The crux of the problem for Western security forces is the asymmetric nature of cyber warfare. It is much easier and cheaper to launch a cyber-attack than it is to defend against one. The attacker must only identify a single point of failure and manipulate it. The defender must protect against every possible type of attack. This works to the advantage of rogue actors, who are able to project more power through cyber warfare than they could through traditional warfare. "A cyber battle would be a lot riskier for the United States and its allies than for North Korea," explained Robert Silvers, former assistant secretary for cyber-policy at the Department of Homeland Security, under President Obama.

"Cyber warfare is simply the modern chapter in North Korea's long history of asymmetrical warfare," concluded a security research report published by HP in August 2014. "Cyber warfare provides [North

Korea] a strategic advantage since outbound attacks are possible, but inbound attacks would have limited reach." Cyber warfare, Recode reported, allows North Korea to offset military weakness on the ground in the air.[124] "Security experts and former intelligence officials believe that battlefield favors North Korea," Wired reported.

North Korea's backwardness is an advantage for the country, because it creates few touch points that American hackers can take advantage of. North Korea has just 1500 IP addresses.[125] "Little of the Hermit Kingdom's operations [that] do connect to the internet are likely as vulnerable to US hacking," Wired continued, noting that North Korea's Internet is wholly independent of its Intranet, where more sensitive information is stored. For these reasons, Richard Clarke, a cyber-intelligence expert, ranked North Korea as the #1 most prepared for a cyber conflict in his 2010 book *Cyberwar*. The United States, with its freedom and prevalent Internet access (with the Internet of Things, even our refrigerators will be connected to the Internet), ranked dead last for cyber-conflicted preparedness.

The U.S. is not weaker than North Korea. Our hackers are better, not worse. Certainly, they are better trained, more professional and more potent. But so long as North Korea is on the offense, and the United States is on the defense, we will always be at a strategic disadvantage. Because cyber warfare, by its very nature, favors the attacker. Therefore — amateurs and teens have been remarkably successful hackers. With a simple Internet connection, and a toolbox you can read about on online forums, any young person can become a potent hacker, and create thorny problems for U.S. companies, and even the government. In America, where hackers can be arrested, and the penalties are substantial, the legal system has helped deter bad behavior. In North Korea, where attackers are rewarded, rather than punished with jail time, you can understand why hacking has become so prevalent and difficult to stop.

Coming back to SONY, a postmortem review helps us understand why the NSA missed the warning signals and failed to prevent the attack. A *Times* investigation found that the NSA's technology would have allowed them to see the "spear phishing" attacks against SONY, where targeted emails are sent to steal login credentials from specific users and gain access to the company network. Somehow this "didn't set off alarm bells," according to the *Times*. Investigators believe that North Korean hackers spent at least two months, from mid-September to mid-November, mapping Sony's computer infrastructure, finding sensitive filings, and plotting how they would destroy key servers and computers.

"They were incredibly careful, and patient," said one person briefed on the investigation told the *Times*. Even with an inside view of North Korea activities, the NSA "couldn't really understand the severity" of planned attacks. So much did the U.S. underestimate the significance of these early warning signs, that when General James Clapper, Director of National Intelligence, met with his North Korean counterpart in early November, he made no mention of the Sony hacking campaigns. He did not demand it stop — or inquire about the intelligence he had received (if he was briefed at all on the topic).

Even as the sophistication of the Sony attack shocked the U.S. intelligence community (to the point that some officials claimed it was too sophisticated to have been carried out by North Korea alone), other had been intensely aware of the North's growing power and skills. Jang Sae-yul, a former North Korean army programmer and defector, reported that the North Koreans "Have built up formidable hacking skills. They have spent almost 30 years getting ready, learning how to do this and this alone, how to target specific countries."

With this context in mind, let us turn now to our core question: What threat do North Korean hackers — who are clearly sophisticated — pose to U.S. elections. We believe the threat is serious. To date, there is little evidence that North Korea has leveraged its cyber warfare capabilities to attack U.S. elections. (Nonetheless, the Trump administration went on a media blitz following his victory, claiming that any election improprieties were the responsibility of North Korea, not Russia). Whether or not North Korea has hacked our elections in the past, there is little reason to feel secure. If North Korea has failed to manipulate U.S. elections, it is only because they haven't tried yet. Given the sorry state of U.S. voting machine security today, North Korea's 6,000-man strong hacker corps poses a serious threat to the integrity of our democracy. We should not ignore it.

Iran's cyber warfare capabilities are remarkably similar to North Korea's. The two countries have much in common: both are rogue states, sanctioned by the west, both have nuclear ambitions, and both blame Western forces for their economic woes. Iran is a religious dictatorship; North Korea is a suppressive regime run by a family dictatorship. The countries are believed to have shared information in pursuit of their nuclear programs and are said to collaborate on malware development. At the very least, the countries study each other's attacks, with the goal of replicating them. In 2012, for example, Iran launched a virus that wiped the memory of 30,000 computers operated by Saudi Aramco, and 10,000 servers. The data on these machines was replaced with an image of a burning American flag. Seven months later, according to the *New York Times,* a similar piece of malware was used to attack three South Korean banks and two broadcasters. Like North Korea, Iran benefits from the asymmetric nature of cyber warfare. Unlike North Korea, Iran has focused its cyber warfare capabilities almost exclusively on achieving its geopolitical aims in the Middle East. Thus, Iran poses less of threat to the United States in the short-run. In the long-run, its advanced cyber warfare capabilities have the potential to cause serious damage to the United States and our elections.

Iran is a sophisticated cyber power. Though it developed its cyber warfare capabilities more slowly than either Russia or China, its technology is potent. Iran's hackers began their malicious campaigns in the early 2000's, when they began taking over websites and "defacing" them with pro-Iran propaganda. The Iran Hackers Sabotage group, for example, launched with the goals of "showing the world that Iranian hackers have something to say in the worldwide security." The group claimed to provide vulnerability tests and web hosting services — standard fare for cyber security companies — but also used its manpower for offensive purposes.[126]

Nonetheless, the "defacers" of the 2000s were relatively benign. Even though they were a nuisance for web administrators, they were nothing more than online protesters, albeit, using illegal means to spread their messages. In 2005, Iran Hackers Sabotage replaced the Guantanamo Naval Station home page with a page condemning terrorism and defending Muslims. By the 2008, the group was thought to be responsible for 3,763 attacks, according to *Newsweek*. Likewise, another group, the Ashiyane Digital Security Team, provided free hacking tutorials, and launched tens of thousands of web defacements, while also providing standard security services. The last of the pseudo-independent Iranian hacker groups to form was Iran's Cyber Army (ICA), which is thought to operate on behalf of Iran's Islamic Revolutionary Guards, which is a unit of Iran's military and a group with widespread influence in the country. The Revolutionary Guard itself employs more than 2,400 hackers, while supporting independent hacking organizations, like ICA.

In 2012, the nature of Iran's cyber capabilities began to evolve, beginning with Supreme Leader Ayatollah Khameini's creation of the Supreme Council of Cyberspace whose goal was to defend Iran against foreign attacks, and lead the way in infiltrating enemy computers. The escalation was believed to be a response to the Stuxnet attack, which targeted Iranian centrifuges, took out 1,000 of the 5,000 centrifuges and set back its nuclear program. By 2014, Gabi Siboni, a cyber security expert with Israel's Institute for National Security Studies, declared "Iran should be considered a first-tier cyber power." Over the course of a two-year period, Iran evolved from being a country of amateur hackers defacing websites for fun, to one capable of advanced cyber warfare, hijacking websites and destroying data.[127]

The Cutting Sword for Justice, for example, launched a massive attack against Saudi Aramco. The attack was sophisticated: it used a virus, called Shamoon, to wipe more than 30,000 computers. The attackers claimed to be protesting "Saudi oppression and corruption, financed by oil." In 2014, a similar attack was launched against the Las Vegas Sands Corporation, whose largest shareholder, Sheldon Adelson, proposed that the U.S. drop a bomb on the Iranian desert to dissuade the country from its nuclear ambitions. In 2015, Iranian hackers penetrated a Navy and Marine Corps computer network, which officials say took four months to recover from. In 2016, the Shamoon virus was used against Saudi Arabia's aviation agency.

Distributed denial of service attacks (DDOS), the tool of choice for many amateur hackers, is also a popular choice for Iranian hackers, both independent and government-sponsored. DDOS attacks overload a website with Internet traffic and thus make it inoperable. In 2012 and 2013, the Cyber Fighters of Izz ad-Din al-Qassam attacked major U.S. banks, which it called "properties of American-Zionist Capitalists." (Fun aside: In the Final Round of the Henry Hudson Debate Tournament in 2013, your authors claimed that Iran's nuclear capabilities were a bigger threat than Iran's cyber warfare capabilities. We were defeated in a 3-2 decision among the five judges. The majority of the panel agreed that a sophisticated attack on the U.S. financial system could be as devastating as a nuclear war, and certainly is more likely).

Iran's financial attacks caused tens of millions of dollars in damage, according to a U.S. law suit, in which seven Iranian hackers were indicted, in absentia.[128] Including in the indictments was the claim than one of the attackers hacked into the Bowman Avenue Dam's infrastructure, which would have allowed him to "operate and manipulate" the dam's gates. (Fortunately, the dam was offline for maintenance at the time). Meanwhile, Advanced Persistent Threat 33, another cyber warfare team, has attacked petrochemical, defense and aviation companies. Advanced Persistent Threat 34 targets financial, energy, telecommunication and chemical companies.

Time Magazine honed-in on the Bowman Dam attack in a profile: "Iranian Cyber Attack on New York Dam Shows Future of War." The attackers — who made their way to the FBI's "Wanted" list, used a cellular modem to gain access to the dam's command and control system. The attack scared U.S. policy makers, because it foreboded a dangerous attack scenario, where our enemies could control our infrastructure, in the case of a conflict. "They were sending a shot across our bow," New York Senator Charles Schumer told *Time*. "They were saying that we can damage, seriously damage, our critical infrastructure and put the lives and property of people at risk."[129]

"The infiltration of the Bowman Avenue dam represents a frightening new frontier in cybercrime," explained U.S. Attorney Preet Bharara of the Southern District of New York. "These were no ordinary crimes, but calculated attacks by groups with ties to Iran's Islamic Revolutionary Guard and designed specifically to harm America and its people."

How did Iran develop such sophisticated capabilities? The leap from defacing websites to sophisticated malware is believed to have been achieved with the help of foreign powers, according to *Newsweek*. Former Congressman Peter Hoekstra and Matthew McInnis, a resident fellow at the American Enterprise Institute were reported to have attributed Iran's cyber warfare capabilities to the country's ties with Russia.

"Containing Iran's cyber warfare program would likely be even more challenging than containing its nuclear program," concluded Dorothy Denning in her December 2017 report for *Newsweek*. Denning is a prominent security researcher and Professor of Information Warfare and Security at the Naval Postgraduate School. "Computer code is easy to conceal, copy and distribute, making it extremely difficult to enforce controls placed on cyber weapons. That leaves cyber security and cyber deterrence as America's best options for defending against the Iranian cyber threat."

Yet not everyone agrees with Denning's conclusion. Indeed, MalCrawler, a cyber security group that protects industrial clients, told the *Financial Times* that Iran does not respond to the same deterrence as traditional cyber powers. Whereas Russia and China are focused on creating digital backdoors and stealing IP, both countries respond to deterrence, creating a "digital equilibrium." Not so for Iran. "In the Middle East, that's not the case at all," explained Dewan Chowdhury, who is the CEO of MalCrawler. "The mindset just seemed completely different — it wasn't espionage or some kind of targeted operation necessarily, it was just to do as much damage as possible." (MalCrawler's defenses include setting up fake networks and computer systems, to attract attackers and study their attack patterns and motivations). The result, according to *the Financial Times,* is that "Iran is rapidly emerging as the sixth member of the cyber superpower club," where it joins the U.S., U.K., Russia, China and Israel.[130]

Analysts believe that Iran is more focused than ever on its cyberwarfare capabilities, in the aftermath of its nuclear deal with the United States. "Before the [nuclear] deal, cyber was just one option they used for leverage, but now, post deal, it is even more central to their toolkit," a senior Middle Eastern intelligence official told *FT*. "Iran is poised to do something in cyber that will change the way the world looks at it… the US knows this. [The US] saw what they [Iran] did during the agreement and they know what they are doing after it."

MalCrawler and others describe Iran as unique, because its cyber-attacks tend to be the most destructive. "There is certainly a big move towards having more destructive capability. They want to be able to do more Aramcos. Right now, they are researching, practicing," explained a senior intelligence official. Iran is thought to spend $1Bn a year on its cyber warfare capabilities, half of the U.K.'s budget. New attack groups are gaining prominence. Researchers say the Saudi Aramco attack and the U.S. financial system attacks were merely "the first shot across the bow." Since 2014, the Rocket Kittens have attracted the most attention because it developed customized malware to attack Iranian dissidents and Israeli organizations. Cylance, a leading U.S. cyber security company, has dubbed Iran "the new China" because of the country's obsessive focus with attacking critical infrastructure.

"What Cleaver really brought to the surface was that these guys were aggressive, compromising critical infrastructure in missions that did not have any classic espionage outcome... the Iranians aren't getting into airports and oil and gas companies for intelligence collection... these are systems to compromise in order to do harm," explained John Hultquist, an analyst at cybersecurity group iSight. "What was really eye-opening is that they were doing it globally."

Iran is not only an attacker, analysts say, but is also a proliferator, distributing malware to anti-American groups, like Hezbollah. Iran's Basij Cyber Council also motivates young computer scientists and encourages them to work as "hacktivists." Hacktivists, who are less sophisticated, are used to promote Iranian propaganda on the web. These hackers still focus on simplistic attacks like defacement and are difficult to defend against because they are decentralized and disorganized. "When it comes to supporting private cyber-actors, Iran appears to be one of the worst offenders," explains *Tower Magazine*.[131]

Iran, which has used terrorist proxies to achieve its foreign policy agenda in the past, with its support of Hamas and Hezbollah, is merely expanding this policy to the Internet, analysts believe. In Lebanon, Yemen and Syria, cyber warfare groups are believed to receive training and support from Iran. These groups include Hezbollah, which has used malware to attack Israeli businesses, the Syrian Electronic Army (SEA), which supports the Assad regime in Syria, and the Yemen Cyber Army, which supports the rebel Houthi militia and has launched attacks against Saudi Arabia. The Assad regime, Hezbollah, and the Houthi militia are all Iranian allies. Iran has transformed cyber warfare by proxy into a potent arrow in its foreign policy quiver.

Iranian aggression has been met with a stern response from the United States. Indeed, the U.S. has launched numerous cyber-attacks against Iran, beginning with its famous Stuxnet virus in 2012. "The United States has repeatedly used cyber weapons to cripple another country's infrastructure, achieving, with computer code, what until then could be accomplished only by bombing a country or sending in agents to plant explosives," the *New York Times* explained, describing U.S. cyber offense against Iran. U.S. offense, aided by Israel, has made the Iranian conflict the closest we've seen to a full-fledged cyber war, with both parties engaged offensively. Unfortunately, because of the asymmetric nature of cyber war, Iran has been neither blocked nor deterred. Over time, it has only scaled up its attacks, and continues to be a serious threat to U.S. national security and election security.[132]

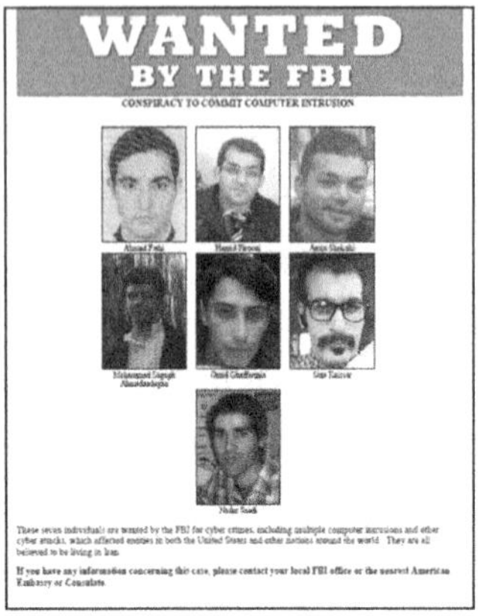

Figure 43: *Iranian hackers make the FBI's most-wanted list*

In 2018, with Iranian protesters again taking to the streets as they did in 2009, we witnessed once again the role that cyber warfare can play, domestically, not just globally. As Dennis Santiago noted for the Huffington Post in his piece "Cyber Warfare Iranian Style" the battle for control of the domestic protests, and indeed, an Iranian revolution, is taking place in cyberspace, where the Iranian government attempts to subvert dissidents, who are using Virtual Public Networks (VPNs) to avoid government control and intervention. As Iran's domestic troubles distract it from its anti-Western campaigns, the protests of 2018 nonetheless reinforce the idea that for Iran, cyber is the new battleground.[133]

Like North Korea, Iran is not believed to be involved in hacking U.S. elections. It is far too busy taking down local political dissidents, spreading its power locally through cyber warfare proxies, and infecting critical infrastructure with malware. Iran is focused on a unique set of geopolitical goals, and U.S. elections don't appear to be a priority. Since both Democrats and Republicans appear to be universally "anti-Iran" (albeit, to different degrees and in different manifestations), Iran may have less to gain from manipulating U.S. elections than a country like

Russia, with its close ties to Donald Trump. Nonetheless, the rising power of Iran is concerning. Should Iran and its proxies target U.S. elections, it would pose a very credible threat.

Taken together, the history of North Korea and Iran's nuclear programs should teach us one clear lesson: despite being isolated and impoverished, these countries are fully capable of launching devastating attacks against Western institutions and infrastructure. Because of the asymmetric nature of cyber warfare, North Korea and Iran have been able to bypass our defenses, taking down major corporations (like Sony) and causing real damage with WannaCry, by taking down hospital computer systems. If countries decide to attack U.S. elections, we'll have a hell of a time defending against them.

Amateurs and Black Hat Hackers

The final two groups of hackers we need to be concerned with are black hat hackers, who often work as hackers for hire, working for profit, and amateur hackers, often self-taught programmers who hack for fun or to achieve a political objective. Though the media paints hackers as sophisticated, well-trained programmers, the reality is that many hackers have limited capabilities, which they have learned on the Internet. However, the proliferations of hacking tools and "how-to" websites have increasingly made these hackers potent. Your authors, for example, would certainly be classified as amateurs. Critically, amateur hackers and black hat hackers have power in their masses and anonymity, making them important threat actors in our election security model.

In the past decade, the hacking industry has developed from an underground, black market economy into a well-understood, modern marketplace of ideas. The rise of black hat hackers — hackers whose goal is to steal information or money for political or economic reasons — has been matched by a rise in white hat, or ethical hackers, to combat them. Today, the rapid rise of the Internet has created a gold rush, for criminal hackers who hope to earn a fortune by taking advantage of weak corporate defenses, and for the cyber security firms and individuals, whose job it is to fight against these criminals. The result is $75Bn+ industry that has invented itself almost overnight.[134]

Who are hackers and where do they come from? Hackers are often computer science professionals or young computer-savvy individuals who are self-taught. Kroll, a cyber security company, segments hackers into three groups: white hat hackers, who are benign and work in the interest of promoting safety on the Internet. White hat hackers are often community oriented and commit to open source projects. Black hat hackers are "bad hackers." Their goal is to maximize damage to a target. Finally, grey hat hackers are a mix of the two; while they have no criminal intent, they have no qualms about breaking the law. Grey hat hackers can be political activists, and thus are extremely relevant to election security. Anonymous and Lulzsec are two famous Hacktivist groups, who use their talent to express political messages. For our purposes, both groups are grey hats.[135]

The old image of hackers being lonely teenage boys is probably outdated. The hacking community is well established. Hacker conferences — like RSA, Redhat, DEFCON and others are fun, community outings. There is camaraderie among hackers, and the community has grown as instant messenger and forums made it easier for hackers to communicate. "The initial lure of hacking" explains *The Guardian*, is "getting past the security hurdles on computers that are intended to turn the vast majority of people away…" is "simply the achievement." Amateur hackers get their start when they realize that while many web targets are secure, there is "long tail" of websites that are trivial to hack. 50% of all computers run the Windows XP operating system, which has known security vulnerabilities. Computers running pirated version of XP don't even have Microsoft patches, intended to fix some of the most egregious vulnerabilities.

As a result, it doesn't take much to get started as a hacker. Some hackers stop here. The simplest exploit is believed to be defacing websites. But as hackers build confidence, many can grow more ambitious. And as the challenges get more serious, so to do the consequences, including jail time. The Digital Millennium Copyright Act (DMCA) bans most forms of hacking. The DMCA makes it illegal to bypass security protections and gain unintended access to a device. DMCA violations can result in more than 10 years of jail time. For some amateurs, however, legal repercussion are not a sufficient deterrent. Hackers are motivated by "the fact that machines will do what you tell them, without argument, again and again. Once mastered, it's a delicious power," says *The Guardian.*[136]

Thousands of programmers, attracted by the challenge and thrill of hacking, now form the core of a growing community. Most of these hackers are relatively benign. For example, Anonymous, a hacktivist group, recruited "Noob" or amateur hackers by giving them guides to participate in Distributed Denial of Service attacks, and to run Twitter bots to uncover ISIS recruiters. "Instead of sitting idle in the channel or lurking around and doing nothing, you can benefit greatly from the different tools and guides that have been provided to you," one Anonymous member said, describing the public toolkit it has made available.[137]

The tool of choice for amateur hacktivists, says the *New York Times*, is the Low Orbit Ion Cannon (LOIC), which is an open-source software program that allows the user to join a botnet, or a connected group of computers, commanded by Anonymous. The combined computing power of all the amateurs allows Anonymous and other hacktivist groups to take down mainstream websites with distributed denial of service attacks. LOIC is apparently widely used — it has been downloaded more than 50,000 times — despite being illegal.[138] LOIC demonstrates just how unsophisticated many so-called hackers are. All these hackers do is download a software application and run it. No coding required.

Nonetheless, amateur hackers can have potentially devastating impacts on the sites they attack. In October of 2016, The *Washington Post* reported that a massive denial of service attack which either slowed or took down Amazon, Twitter, Spotify, and others was in fact the work of amateurs, not the work of a state-actor. "The technical and social indicators of this attack align more closely with attacks from the [HackForums] community than the other type of actors that may be involved, such as higher-tier criminal actors, hacktivists, nation-states, and terrorist groups," explained researchers at Flashpoint, a security firm. [139]

The code for the attack malware, called Mirai, had been posted online to hacker forum at least a month before the final attack. Based on this information, Mikko Hypponen, chief research officer at cyber security firm F-Secure, agreed with Flashpoint's assessment. "I think they are right. I don't believe the Friday attackers were financially or politically motivated," he said. "It was such an untargeted attack, it's hard to find a good motive for it. So, kids." In December 2017, the creators of the Mirai botnet plead guilty in an Alaska court; its creators were all aged 20-22, and the ringleader was a Rutgers University student.

"Once they built the botnet, they sought to make money by renting it or extorting companies for money," explained William Fitzpatrick, the acting U.S. Attorney in New Jersey, where the botnet repeatedly hit the computer network at Rutgers University. Other hackers used the code to launch the distributed denial of service attack that targeted websites including Netflix, Amazon, Twitter and others. The Mirai botnet demonstrates the extent of the damage that can be caused by a group of talented hackers, no matter how small. An attorney for Paras Jha, who was one of the Mirai conspirators, described him as "a brilliant young man whose intellect and technical skills far exceeded his emotional maturity."[140]

Despite their growing prevalence, we do not believe amateur hackers pose a material threat to election security. The reason is simple: whereas computer code is scalable — one program can be run by thousands of people on the Internet and can cause real damage — hacking a voting machine requires a disproportionate amount of physical intervention. You need to physically install malware on the voting machine or manipulate the memory chip. The hacker must be acutely aware that his actions are illegal. There is no plausible deniability. There is no ability to mask the act as "just code." The attack cannot simply be launched from a computer — it must be executed from the voting booth. Not only is the potential to get caught substantially higher for voting machine exploits, but also the ability to scale is very limited. It would be difficult for a single hacker to infect more than a few voting machines, or at best a single county.

We should not discount amateur hackers in our threat model. A group of amateurs could potentially infect voting machines across a few strategic voting districts to influence the outcome of an election. However, we believe state-actors are a much more substantial threat to U.S. elections. State actors have much greater incentives to hack elections, better manpower and coordination, and far superior funding. Espionage is not new for government...it probably is for amateur hackers, who believe they have little chance of being caught from behind their computers. Because we believe that hacking an election is as much an act of espionage as it is a cyber security feat, we do not believe amateurs should be the primary concern when developing a threat model to defend voting machines.

Sophisticated hackers, however, pose more of a threat. Black hat hackers — or hackers for hire — tend to be advanced programmers seeking to profit from their skills. Black hat hackers are potentially more threatening because they will be focused on identifying voting machine exploits and selling those hacks to foreign governments. The best black hat hackers are potentially as good as, or better programmers than, the academic researchers cited throughout this book. Their exploits are potentially more sophisticated than many of the exploits we have reviewed thus far. And black hat hackers don't post their exploits in academic journals or have them published in books like this one — so that they can be fixed and addressed. Instead, black hat hackers sell their exploits to the highest bidder. Therefore, we believe black hat hackers are dangerous insofar as they sell vulnerabilities to state-actors, who have the manpower to coordinate a sophisticated attack on U.S. elections.

An expose of the Italian spyware company Hacker Team sheds light on the illicit industry of black hat hackers who sell "zero-day" exploits to the highest bidder. Zero-day exploits are new vulnerabilities that are still unknown to the manufacturer. Hacker Team uses these exploits to install spyware on personal computers of dissidents, companies and target governments. It then sells the exploits and data collected from spyware to the highest bidder, often governments. Hacker Team has been widely criticized because it has sold exploits to oppressive governments, like Iran. The black market for exploits was a "hidden dark alley to anyone but the hackers and sellers who call it home." Now that Hacker Team has itself been hacked — and 400 gigabytes of its data have been posted to the Internet — there is clarity around how the market operates.[141]

For starters, exploits can be expensive. Exploits sought by Hacking Team were being sold for anywhere from $5000 to $500MM, for unique IOS exploits. The emails released in the hack revealed the details of Hacker Team's negotiations — both with sellers, who were often reluctant, because of Hacker Team's nefarious reputation, and directly with hackers, who began working directly for the company to generate exploits, which it could sell. The leak offers insight into how the exploit market works, but there is still much that is unknown. "Hacking Team is a second-rate company that had to work hard to find people who weren't going to treat it as such," explained Vlad Tsyrklevich, a security researcher who reviewed the documents.

Alongside Hacker Team's zero-day exploits, there is also a growing industry of hackers selling their services for a profit. Hacker's List is one example of such a development, where individuals request the services of hackers and hackers bid on the proposal. Initial projects on the website, which launched in November 2014, ranged from $100 — $5,000. "I want the client lists from a competitor's database. I want to know who their customers are, and how much they are charging them," wrote one Hacker List customer. The websites demonstrate the extent to which hacking has become mainstream, and the barriers to entry are increasingly being lowered for non-technical individuals, hoping to launch cyber-attacks.[142]

"Hackers for hire can permit nontechnical individuals to launch cyber attacks with a degree of deniability, lowering the barriers to entry for online crime," FTI Consulting's Thomas Brown told the *New York Times*. Since advanced U.S.-based programmers already have many lucrative opportunities to work as ethical hackers in the U.S., or for large tech companies, Hacker List and other sites appeal primarily to international hackers, who may have few professional opportunities. The professionalization of the illegal hacker market is concerning because it provides politically motivated individuals an ability to more easily recruit the talent needed to promote their agenda.

We believe the primary threat that black hat hackers pose is as boots on the ground for state-actors. Recruiting hackers, as we've seen, has always been a key challenge for state actors and one of the primary roadblocks to launching a sophisticated attack. While governments will continue to rely on well-trained and reliable government hackers and spies, the ability to hire talent globally increases the potency and reach of state-actors. It makes them more dangerous and increases their ability to project their power and achieve foreign policy goals. In our case, this means the rise of black hat hackers will empower foreign governments and make it easier to hack our elections in the future.

In summary, our threat model for election security implicates dozens of potential threat actors as potential attackers. State-actors like Russia, China, North Korea and Iran continue to pose the most substantial threat to election security, since they are well-funded, motivated to influence U.S. elections and thus the geopolitical order, and have the espionage capabilities and scale to carry out a widespread attack, which could materially affect the outcome of an election. Sitting on top of this core threat, we have amateurs and black hat hackers, who alone are unlikely to launch a sophisticated attack on U.S. elections, but add leverage to the system, making state actors potentially more dangerous and more capable of circumventing voting machine defenses. Together, these threat actors give us more than enough reason to be worried that the voting machine vulnerabilities we have outlined *will* be exploited. Thus, it is more urgent than ever that we update our voting machines to make them more secure. We need a complete overhaul of the current election security model to ensure the integrity of our democracy.

CHAPTER 8

2016, 2018 AND BEYOND

"IP addresses can be invented — a child can do that! Your underage daughter could do that." — Vladimir Putin, Interview with Megan Kelly on Election Hacking

Russia intervened to manipulate the 2016 election. And they'll be back in 2018. Let's start with a timeline of what we know.[143],[144]

September 2015: The FBI notified the Democratic National Committee (DNC) that at least one DNC computer had been hacked by Russia. The DNC investigates the claim, but finds nothing unusual

November 2015: The FBI notifies the DNC that it has been compromised, and that Russia now has access to data on its networks. The IT team does not pass this information along to DNC leadership.

On March 19, 2016, John Podesta, Clinton's campaign boss, clicks on a phishing email. After receiving the go ahead from his IT help desk (apparently, the IT person intended to type, "this is illegitimate" but instead wrote "this is legitimate"), Podesta clicks through and gives his password to Russian hackers.

On June 12, WikiLeaks founder Julian Assange announces he will publish Clinton emails, likely associated with the Podesta hack.

Between October and November of 2017, during the heat of the election, Assange releases 58,000 Podesta emails.

On June 14, *The Washington Post* reports that Russian hackers stole opposition research on Donald Trump, staffer emails, and chat exchanges. Russia denies responsibility.

On June 15, the DNC's cyber security contractor announces that Russian hacker groups known as "Cozy Bear" and "Fancy Bear" were responsible for the attack. At the same time, a blogger known as Guccifer 2.0 claims responsibility for the attack, and posts DNC memos and opposition research as proof.

On June 22, WikiLeads publishes 20,000 emails hacked from the DNC. DNC Chair Debbie Wasserman Schultz resigns of evidence that she may have been biased in favor of Hilary Clinton and against Bernie Sanders.

On June 25, the FBI announces it is investigating the DNC hack.

In August 2016, President Obama received a report explained that Vladimir Putin was going to disrupt the 2016 election using cyber warfare. President Obama begins to investigate voting machine vulnerabilities and countermeasures.

On August 3, Homeland Security Secretary Jeh Johnson announced that voting machines might be "critical infrastructure" a designation that would allow Homeland Security to provide funding and intelligence to local election officials.

On August 5, the National Association of Secretaries of State announced, "Hacking of the election is highly improbable due to our unique, decentralized process." Essentially, they refused to acknowledge their own vulnerabilities.

On August 12, hackers release Nancy Pelosi's cell phone numbers and email address, among other information. Pelosi describes receiving "obscene and sick calls."

On August 15, Jeh Johnson makes an official offer of assistance with elections. For the most part, states do not accept aid, because they do not want federal interference in their right to manage elections.

On August 18, the FBI releases a list of IP addresses with known hacker associations. States check for intrusions from these IP addresses. Once the tally is complete, 21 states find that hackers targeted them.

On August 24, Russian hackers attacked VR Systems, a software vendor that provides the voter registration system used in local elections. The attack is launched using a "spear phishing" campaign, where employees receive targeted emails, attempting to convince them to reveal their login credentials, or install malware on their computers. VR Systems claims none of the spear phishing emails succeeded. The NSA concludes otherwise, stating that it believes at least one employee account was compromised.

On August 31, the Department of Homeland Security establishes the Election Infrastructure Cybersecurity Working Group, to share information among the states and advice on defense measures.

On September 1, Putin tells Bloomberg that Russia is not behind the hacks. Nonetheless, he says the hacks were beneficial since they helped inform U.S. voters.

On September 22, Senators Dianne Feinstein and Adam Schiff announce that they believe Russia is attempting to interfere in the U.S. election, based on information they received through intelligence briefings.

On September 26, during a presidential debate, Donald Trump says he does not know who is behind the DNC hack. "It could be Russia, but it could also be China. It could also be lots of other people. It also could be somebody sitting on their bed that weighs 400 pounds," he says.

On September 28, House and Senate leaders announced their opposition to DHS efforts, claiming that it was an attempt to violate state rights. "For over 200 years the states have overcome every challenge to ensure the smooth function of our democracy," the senators announced.

On September 30, the FBI holds a conference call with election supervisors to warn them that a local vendor (likely VR Systems) had been hacked.

On November 29, Democratic Senators send President Obama a letter asking him to declassify information linking Russia to election hacks They also confirm that vulnerability scans had "originated from servers operated by a Russian company."

On October 6, DCLeaks, a supposed hacktivist group later linked with Russian military intelligence, publishes documents stolen from Capricia Marshall, a Clinton supporter.

On October 7, Homeland Security and Office of National Intelligence on Election Security officials announce that they are "confident that the Russian Government directed the recent compromises of emails from US persons and institutions." They also confirm that vulnerability scans had "originated from servers operated by a Russian company."

On October 10, the Department of Homeland Security discloses that 33 states and local election agencies sought cyber security advise from DHS. He urges others to follow suite, saying "Time is a factor... there are only 29 days until Election Day."

On October 27, Russian military hackers used a compromised VR Systems email address to send malware to 122 local government offices.

On November 8, elections proceed without any obvious evidence of hacking (some say irregularities and voting machine failures in Texas, Utah, North Carolina, and Pennsylvania could have been associated with hackers).

On December 9, *The Washington Post* reports that the CIA believes Russia meddled in the 2016 election in order to guarantee Trump's victory. Trump denies all claims, saying about the CIA, "These are the same people that said Saddam Hussein had weapons of mass destruction."

On December 10, Senators John McCain, Chuck Schumer, Lindsey Graham and Jack Reed announced that they are working together to stop future cyber attacks.

On December 11, CNN reports that the FBI and CIA disagree on the goal of Russia's election meddling. The CIA believes the goal of the hacking was to help Trump win the election. The FBI believes that goal may have been to cause chaos, more generally, and does not have evidence to support the claim that the hacking was purely pro-Trump.

On December 12, CNN breaks news that Russia also hacked Republican computers and emails. However, the hackers did not release the documents from those hacks.

On December 13, *The New York Times* publishes a feature story explaining what went down during the DNC hack. The *Times* claims that stolen documents were offered to reporters in Congressional districts where Democratic incumbents faced stiff competition.

On December 29, Obama announces a new round of sanctions against Russia for their attempts to hack the 2016 election and expels 35 Russia diplomats.

On January 3, Julian Assange tells Hannity that he did not receive DNC emails from the Russian government.

On January 6, 17 intelligence agencies issue a joint statement saying that Russia attempted to influence the election using "fake news" and by hacking into voting networks. Trump asserts, "There was no tampering whatsoever with voting machines."

On January 6, a declassified report on Russian meddling is released to the public. The report says that voting machines were not hacked, but that Russia meddled in other ways, namely through a propaganda campaign which involved email hacking and fake news.

On January 7, Trump says he recognizes Russia played a role in the DNC hack, but claims that voting machines were not breached. He announces his intent to improve diplomatic relations with Russia.

On March 20, during their testimony in front of the House Intelligence Committee, FBI Director James Comey and National

Security Agency Director Mike Rogers say that they do not believe Russian hackers changed any votes. They acknowledge that Russia will continue meddling in our elections, with Comey telling the committee that "We have to assume they're coming back."

On June 1, Putin announces that "Patriotic citizens" may have been involving in election hacking. Putin says the government had no role in the election hacks. In an interview with Megan Kelly, Putin said that even a child could have hacked the U.S. election.

On June 5, NSA contractor Reality Winner is arrested for leaking a report, which concluded that Russian intelligence officials were behind the VR Systems hack. There is uproar that states had not been briefed with this information.

On June 21, Samuel Liles, a Homeland Security Official, tells the Senate Intelligence Committee that Russia targeted elections in 21 states. Liles claims that no votes were changed. At this point, states had not been notified about whether they were targeted.

On June 25, Congress passes a bipartisan bill, which prevents Trump from lifting Russian sanctions. The bill passes 419-3.

On June 30, Putin expels 755 employees in Russian diplomatic missions, in retaliation against U.S. sanctions.

On July 30, Trump signs the Russia sanctions bill, despite calling it "seriously flawed."

On September 6, Facebook announces that 3,000 ads on its website between June 2015 and May 2017 were linked to Russia.

The Washington Post reports that the ads were linked to Internet Research Agency, a Russian company.

On September 22, Homeland Security notifies 21 states that their election infrastructure had been hacked, including networks that contained voter information.

On October 3, CNN reports that the Russian Facebook ads were specifically used to target Michigan and Wisconsin, swing states that Trump won narrowly.

On October 12, CNN republishes an expose about a group of Russian "trolls" who claimed to be Black Lives Matter activists during the campaign to influence voters.

There has been much debate around Russia's role in the 2016 election. We want no part in that debate. The facts are the facts, and we can stop there. As the timeline above illustrates, the facts are fairly simple: Russia meddled in the 2016 election. The extent of the damage they caused — and whether Trump would have won without their meddling — is difficult to quantify. It's an impossible task, with little to be gained from the exercise. Instead, we should be focused on 2018 and 2020. If Russia meddled in 2016 — to great fanfare and effect — it is likely they will try again in future elections. China, North Korea and Iran might be inspired too. Black hat hackers like Guccifer 2.0 will surely join in.

Our elections are not secure. On top of their campaign for misinformation, Russia may have hacked voting machines in 2016. If they did, we would really have no way of knowing. "As a remote attacker, I can target an election management system, one of these ballot programming computers," cyber security researcher Alex Halderman told NPR. "If I can infect it with malicious software, I can have that malicious software spread to the individual machines on the memory cards, and then change votes on Election Day." Because 14 states have no paper ballot backups — and those states that do have paper backups rarely audit them, such an attack would not be detected. Reporting on known Russian hacks just barely scratch the surface of what hackers are capable of.

"This is a real threat," the Brennan Center's Larry Norden told NPR. "It's not going away, and if anything, foreign adversaries, even people at home, might be emboldened to do this more going forward. And to me it is a real call that we have to do more as soon as possible to secure these systems." Connecticut Election Director Peggy Reeves told the National Academies of Sciences, Engineering, and Medicine that as of June 2017, little had changed in terms of election security. "Many of our towns actually have no local IT support," she said. "Seriously, they don't have an IT director in their town. They might have a consultant that they call on if they have an issue. So they look to us, but we're a pretty small division." Reeves correctly notes that our best defense is that elections are decentralized. Increasingly, it looks like that might not be enough, in the way of protecting our elections. [145]

So of course, we should be worried about the future. And if anything, this book provides ample evidence of the reasons we should be concerned. Hackers made a joke out of our election security in 2016. With so many vectors to attack voting machines in 2018, hackers may start changing vote counts. Over time, they will only get more sophisticated, harder to track, and harder to detect. Access control vulnerabilities, voting machine viruses, and data overwrite / cryptography vulnerabilities offer hackers three vectors of attack. And the threat actors are numerous. Russia, China, Iran, and North Korea have sophisticated hacking capabilities. Russia made a concrete effort to influence the 2016 election. They might try again in 2018. Black hat hackers and amateurs working for the wrong side add fuel to fire.

Countermeasures abound, but none have been implemented. Voting machine viruses, voter verified ballots, open source voting machines, enhanced cryptographic protections, improved physical security, ensuring paper ballot back-ups are just some of the countermeasures that have been proposed. There is no shortage of solutions to our election security crisis. Clearly, the chorus of technical voices saying, "fix encryption," "fix physical security," "use open source voting machines" is not solving the problem. Professors at Princeton, Michigan, Penn and other elite universities have been vocal about the need for reform for more than a decade, and they have been ignored.

Today, we have a unique opportunity to take what has been thus far been a movement among technical cyber security professionals and make this a national priority. Indeed, for the first time ever, everyday citizens are concerned with election security. Voters are demanding reform. There is a bipartisan mandate to shake up our elections — just like there was back in 2000, when Bush v. Gore exposed flaws in punch card voting systems. We need to capture this momentum and channel it into tangible reform. There is the political will to reform elections, if only we can translate it into real legislation. Stricter standards for voting machine manufacturers and better physical security for voting machines are easy places to start.

We need to create a political movement that brings the impetus for change to the gates of Washington, our statehouses, and our local governments. If every local government responsible for buying voting machines refused to by voting machines that did not meet a certain security standard, they would force manufacturers to step up their game. We, as voters, must send a message to public officials that we will no longer tolerate complacency. That we value our right to vote and demand that the government protect that right with adequate security procedures. And the media needs to keep up the pressure. By reporting on election improprieties, the media ensures that voters do not forget about the important threat challenging our elections.

Despite the uproar over Russia's meddling in the 2016 election, no action has been taken. In Detroit, where voting machine failures were the most dramatic (and potentially influential, because Detroit is heavily Democratic as compared with the rest of Michigan, which turned out for Trump), election officials decided to order a new batch of voting machines from the same bad actors, the same manufacturers who leave hardcoded encryption keys in their code bases. This is not a real solution. In the short run, it may make the situation slightly more tolerable — the worst and most obvious bugs will be fixed in new voting machines — but in the long run, more bugs and vulnerabilities will be found. Voting machine manufacturers have little incentive to invest in security. In San Francisco and Los Angeles, open source voting machines are gaining traction, and will be in use by 2020. This is a step in the right direction. If these cities can show that open source voting machines are workable solutions, then others might follow suit.

In Washington, reformers exist, but their pleas for change have fallen on deaf ears. Senators Amy Klobuchar (D-MN) and James Lankford (R-OK) have proposed a series of reforms to protect states in future elections, including enhanced information sharing, back-up paper ballots, and funding for states to implement vulnerability testing. Klobuchar and Lankford are focused on ensuring that elections can be audited — today 12 states have absolutely no ability to audit their elections. Their bipartisan initiative is admirable, but outside of their ten minutes of fame on PBS News Hour, their initiative has largely been ignored. Still, 48 states have not updated their election systems in more than a decade.[146]

The worst is this: even if every one of Lankford and Klobuchar's reform proposals were implemented, the 2018 election will still be hacked. The clock has run out on election reform before the next election. In the two years since America confronted the worst foreign meddling in its history, nothing has changed. 2018 could be a repeat of 2016. There is time to replace election infrastructure and implement new security protocols before 2020. For reformers, 2020 will be a crucial test. With Donald Trump up for re-election, we can expect plenty of drama. If nothing changes by 2020, we will have squandered our opportunity to make a difference. That would be a real shame.

We hope you have enjoyed this journey into the trenches on voting machine security. We hope this story — of courageous security researchers, dangerous and talented state-sponsored hackers, and perilous voting machine vulnerabilities — has been as entertaining for you to read as it was for us to write. Good luck to you on your studies of election security. May we be fortunate enough to see reform implemented in the near future.

FIGURES CITED

Figure 1: McDaniel et al. (2007)

Figure 2: McDaniel et al. (2007)

Figure 3: Kohno et al. (2004)

Figure 4: Kohno et al. (2004)

Figure 5: Oren, et al. 2009

Figure 6: Lightning Talks Day 1, ttps://media.ccc.de/v/22C3-911-en-lightning_talk_day_1#video&t=1438

Figure 7: Appel, et al. 2009

Figure 8: Butler, et al. 2008

Figure 9: Butler, et al. 2008

Figure 10: ES&S Website

Figure 11: King, Samuel. "Wisconsin recount observers cite broken warranty seals." Fox 9. December 4, 2016.

Figure 12: Appel, et al. 2009

Figure 13: Checkoway, et al. 2009

Figure 14: Appel, et al. 2009

Figure 15: Checkoway, et al. 2009

Figure 16: Checkoway, et al. 2009

Figure 17: Feldman et al. 2006

Figure 18: Feldman et al. 2006

Figure 19: Butler, et al. 2008

Figure 20: Aviv, et al. 2008

Figure 21: Aviv, et al. 2008

Figure 22: Appel et al. 2009

Figure 23: Appel et al. 2009

Figure 24: Appel et al. 2009

Figure 25: Essex, et al. 2012

Figure 26: Essex, et al. 2012

Figure 27: Keller et al. 2005

Figure 28: The Geneva Internet Voting System. Council of Europe. 2010.

Figure 29: The Geneva Internet Voting System. Council of Europe. 2010.

Figure 30-36: Images from AVC Edge Demonstration, by David Cahn

Figure 37-39: Emails Received by David Cahn

Figure 40-42: Taxonomies

Figure 43: Santiago, Denis. "Cyber Warfare Iranian Style." Huffington Post. January 9, 2018.

ENDNOTES

Introduction

1. Friedman, Brad. "On Heels of Diebold/Premier Purchase, Canadian Firm Also Acquires Sequoia, Lies About Chavez-Ties in Announcement." The Brad Blog. June 2010.

2. Sanger, David. "U.S. Officials Defend Integrity of Vote, Despite Hacking Fears." *New York Times*. November 26, 2016.

3. Associated Press. "U.S. Tells 21 States That Hackers Targeted Their Voting Systems." *New York Times*. September 22, 2017

4. Fessler, Pam. "10 Months After Election Day, Feds Tell States More About Russian Hacking Facebook Twitter Flipboard Email." *NPR*. September 22, 2017.

5. Fessler, Pam. "Report: Russia Launched Cyberattack On Voting Vendor Ahead Of Election Listen." *NPR*. June 5, 2017.

6. Naylor, Brian. "Intelligence Chiefs 'Stand More Resolutely' Behind Finding Of Russia Election Hacking Facebook Twitter Flipboard Email." NPR. January 5, 2017.

7. Keller, Arthur, Alan Dechert, Karl Auerbach and David Mertz. "A PC-Based Open-Source Voting Machine with an Accessible Voter-Verifiable Paper Ballot." FREENIX Track: 2005 USENIX Annual Technical Conference. 2005.

8. We have changed Steve's name to preserve his privacy

Chapter 1

9. Tribe, Laurence. "The Constitution in Crisis: From Bush v. Gore to the War on Terrorism. The Tanner Lectures on Human Values." Oxford University.

10. Brace, Kimball. "Almost 55 Million, or One-Third of the Nation's Voters, Will Face New Voting Equipment in 2006 Election." Election Data Services.

11. Wofford, B., Heilbrunn, J., Karni, A., & Gollan, J. "How to Hack an Election in 7 Minutes." Politico. August 5, 2016.

12. Chardy, Alfonso. "Leahy Blames Unskilled Workers." *Miami Herald.* September 13, 2002.

13. Keller, Arthur, Alan Dechert, Karl Auerbach and David Mertz. "A PC-Based Open-Source Voting Machine with an Accessible Voter-Verifiable Paper Ballot." FREENIX Track: 2005 USENIX Annual Technical Conference. 2005.

14. Zetter, Kim. "E-Voting Undermined by Sloppiness." Wired. December 17, 2003.

15. Herdon, Ray. "7,000 Orange County Voters Were Given Bad Ballots." *Los Angeles Times.* March 9, 2004.

16. Housewright, Ed. "Area Democrats say early votes miscounted." *The Dallas Morning News.* October 22, 2002.

17. Downs, Caleb. "Trump, Gov. Abbott weigh in on claims that Texas machines changed Trump votes to Clinton." *Dallas Morning News.* October 26, 2016.

18. Appel, Andrew, Maia Ginsburg, Harri Hursti, Brian W. Kernighan, Christopher D. Richards and Gary Tan. "Insecurities and Inaccuracies of the Sequoia AVC Advantage 9.00H DRE Voting

Machine." Princeton University. August 10, 2009.

19. Zetter, Kim. "Virginia Finally Drops America's 'Worst Voting Machines' Wired. August 17, 2015.

20. Alter, Charlotte. "Detroit Voting Machine Failures Were Widespread on Election Day." *Time*. December 13, 2016.

21. Helms, Matt. "Detroit City Clerk Winfrey: We need younger, tech-savvy poll workers." *Detroit Free Press*. January 27, 2017.

22. Jarvis, Craig. "McCrory calls for recount of Durham votes, cites other 'irregularities.'" *Raleigh News & Observer*. November 12, 2016

23. Dayton, Rick. "Voting Issues: Some Trump Voters Reporting Ballots Switching to Clinton." CBS Local. November 8, 2016.

24. Tanaka, Kevin. "Election 2016: Voters face machine problems, long lines in some states." *Chicago Tribute*. November 8, 2016.

25. Owen, Tess. "Utah vote glitch." Vice News. November 8, 2016.

26. Menn, Joseph. "U.S. election agency breached by hackers after November vote." Reuters. December 15, 2016.

27. Talbot-Zorn, Justin. "How to restore faith in American elections." The Hill. October 31, 2016.

28. Sanger, David. "U.S. Officials Defend Integrity of Vote, Despite Hacking Fears." *The New York Times*. November 26, 2016.

Chapter 2

29. Aviv, Adam, Pavol Cerny, Sandy Clark, Eric Cronin, Gaurav Shah, Micah Sherr, and Matt Blaze. "Security Evaluation of ES&S Voting Machines and Election Management System." USENIX. July 28, 2008.

30. McDaniel, Patrick, Kevin Butler, William Enck, Harri Hursti, Steve McLaughlin, Patrick Traynor, Matt Blaze, Adam Aviv, Pavol Cerny, Sandy Clark, Eric Cronin, Gaurav Shah, Micah Sherr, Giovanni Vigna, Richard Kemmerer, Davide Balzarotti, Greg Banks, Marco Cova, Viktoria Felmetsger, William Robertson, Fredrik Valeur, Joseph Lorenzo Hall, and Laura Quilter. "Evaluation and Validation of Election Related Equipment, Standards and Testing. EVEREST Report." December 14, 2007.

31. HDNet. "Dan Rather Reports to Present Conclusive Evidence of Voting Machine Failures." Business Wire. August 12, 2007.

32. Kohno, Tadayoshi, Adam Stubblefield, Aviel Rubin, Dan Wallach. "Analysis of an Electronic Voting System. IEEE Symposium on Security and Privacy." February 2004.

33. Amazon. "YARONG-10 PCS white PVC IC with SLE 4442 card ISO 7816 blank Smart contact IC card. eBay. Titanium Card 2 - Programmable High Security Access Smartcard."

34. Oren, Yossef and Avishai Wool. "Attacks on RFID-Based Electronic Voting Systems." Tel Aviv University. 2009.

35. O'Connor, Mary. "Alameda County Gives RFID Ballot-Tracking a Vote of Approval." RFID Journal. January 4, 2008.

36. Keerthana, G, P.Priyanka, K.Alise Jenifer, R. Rajadharashini, and Aruna Devi. "Impressive Smart Card Based Electronic Voting System. International Journal of Research in Engineering and Technology." March 2015.

37. G. P. Hancke. "Practical attacks on proximity identification system." In SP '06: Proceedings of the 2006 IEEE Symposium on Security and Privacy, pages 328–333, Oakland, CA, 2006. IEEE Computer Society.

38. Lightning Talks Day 1. https://media.ccc.de/v/22C3-911-en-lightning_talk_day_1#video&t=1438

39. Appel, Andrew, Maia Ginsburg, Harri Hursti, Brian W. Kernighan, Christopher D. Richards, Gang Tan, Penny Venetis. "The New Jersey Voting-machine Lawsuit and the AVC Advantage DRE Voting Machine." USENIX. August 2009.

40. *Gusciora et al. v. Corzine et al.*, Docket No. MER-L-2691-04, Superior Court of New Jersey

41. Proebstel, Elliot, Sean Riddle, Francis Hsu, Justin Cummins, Freddie Oakley, Tom Stanionis, and Matt Bishop. "An Analysis of the Hart Intercivic DAU eSlate." USENIX. June 2007.

42. Butler, Kevin, William Enck, Harri Hursti, Stephen McLaughlin, Patrick Traynor, and Patrick McDaniel. "Systematic Issues in the Hart InterCivic and Premier Voting Systems." USENIX. July 2008.

Chapter 3

43. Aviv, Adam et al *op. cit.*

44. McDaniel, Patrick, *op. cit.*

45. Aviv, Adam et al *op. cit.*

46. McDaniel, Patrick, *op. cit.*

47. Aviv, Adam et al *op. cit.*

48. McDaniel, Patrick, *op. cit.*

49. ES&S Website. http://www.essvote.com/products/3/6/dre

50. King, Samuel. "Wisconsin recount observers cite broken warranty seals. "Fox 9. December 4, 2016.

51. Appel, Andrew et al, *op. cit.*

52. J. A. Halderman and A. J. Feldman. "AVC Advantage: hardware functional specifications." Technical Report TR-816-08, Department of Computer Science, Princeton University, Mar. 2008.

53. Checkoway, Stephen, Ariel J. Feldman, Brian Kantor, J. Alex Halderman, Edward W. Felten, and Hovav Shacham. "Can DREs Provide Long-Lasting Security? The Case of Return-Oriented Programming and the AVC Advantage." USENIX. August, 2009.

54. Feldman, Ariel, J. Alex Halderman, and Edward W. Felten. "Security Analysis of the Diebold AccuVote-TS Voting Machine." USENIX. September 13, 2006.

55. Wofford, B., *op cit.*

56. Butler et al., *op cit.*

Chapter 4

57. Aviv, Adam et al., *op cit.*

58. McDaniel, Patrick, *op cit.*

59. Quist, Danny. "Covert Debugging: Circumventing Software Armoring Techniques. Offensive Computing." 2007.

60. Appel, Andrew et al., *op cit.*

61. Kohno, Tadayoshi et al., *op cit.*

62. B. Schneier. Applied Cryptography: Protocols, Algorithms, and Source Code in C. John Wiley & Sons, New York, second edition, 1996.

Chapter 5

63. Butler et al., *op cit.*

64. Feldman, Ariel et al., *op cit.*

65. DRE analysis for May 2006 primary, Cuyahoga County, Ohio. Available at http://bocc.cuyahogacounty.us/GSC/pdf/esi cuyahoga final.pdf, Aug. 2006.

66. Feldman, et al. *op cit.*

67. *Ibid.*

68. *Ibid.*

69. D. Chaum. "Secret-ballot receipts: True voter-verifiable elections." IEEE Security and Privacy, 2(1):38– 47, 2004.

70. Karlof, C., Sastry, N., and Wagner, D. "Cryptographic voting protocols: A systems perspective." In Proc. 14th USENIX Security Symposium (2005).

71. Essex, Aleksander, Jeremy Clark, Urs Hengartner, and Carlisle Adams. "Eperio: Mitigating Technical Complexity in Cryptographic Election Verification." USENIX. April 3, 2012

72. Zetter, Kim. Aussies Do It Right: E-Voting. Wired. November 3, 2003.

73. Keller et al, *op cit.*

74. ProjectEVM 2003. http://evm2003.sourceforge.net/

75. Babe, Ann. "Can Open-Source Voting Tech Fix The U.S. Elections System?" Forbes. February 6, 2015.

76. McMillan, Robert. "Open Source Voting Machine Reborn After 6-Year War With IRS." Wired. August 6, 2013.

77. McCarthy, Kieren. "San Francisco prepares to open source its voting system software." The Register. February 10, 2016.

78. Qadah, Ghassan. "Electronic Voting Systems: Requirements, design and implementation." Computer Engineering, American University of Sharajah (UAE). Computer Standards & Interfaces. 2007.

79. Myithi, K. "An Efficient Method to Avoid False Voting Using SMS Voting Approach." International Journal of Computer Science and Mobile Computing. 2014.

80. Alvarez, Michael, Thad E. Hall, and Alexander H. Trechsel. "Internet Voting in Comparative Perspective: The Case of Estonia." California Institute of Technology. June 1, 2009

81. The Geneva Internet Voting System. Council of Europe. 2010.

82. Danchev, Dancho. "Study finds the average price for renting a botnet." ZDNet. May 26, 2010.

83. Ahmed, Taha and Mohamed Aborizka. "Secure Biometric E-Voting Scheme." Arab Academy for Science and Technology and Maritime Transport. 2011.

Chapter 6

84. Brace, Kimball. "Almost 55 Million, or One-Third of the Nation's Voters, Will Face New Voting Equipment in 2006 Election." Election Data Services.

85. Keller et al, *op cit.*

86. Sanger, David. "Russian Hackers Acted to Aid Trump in Election, U.S. Says." The New York Times. December 9, 2016

87. Thanks to Daniel Cahn for help editing this video and uploading it to YouTube.

88. Appel, Andrew et al, *op. cit.*

89. Doyle, Chris. "Long lines, registration problems hold up students on Election Day." *The Daily Pennsylvanian.* November 9, 2016

90. Whisler, Karen. "How Penn voted." *The Daily Pennsylvanian.* November 9, 2016

Chapter 7

91. The Conversation. "How computer hacking is becoming Russia's weapon of choice." The Conversation. February 8, 2017.

92. Fidler, Stephen and Demetri Sevastopulo. "Putin rails against US foreign policy." *Financial Times.* February 10, 2007.

93. Medvedev, Dmitry. "The Foreign Policy Concept of the Russian Federation." July 12, 2008.

94. Ward, Carroll. "Russia's Cyber Forces." Defense Tech. May 27, 2008.

95. The Conversation. "How computer hacking is becoming Russia's weapon of choice." The Conversation. February 8, 2017.

96. "A cyber-riot." *The Economist.* May 10, 2007.

97. Swaine, Jon. "Georgia: Russia 'conducting cyber war'" *The Telegraph.* August 11, 2008.

98. McLeary, Paul. "Russia's Winning the Electronic War." *Foreign Policy.* October 21, 2015.

99. The Conversation. "How computer hacking is becoming Russia's weapon of choice." The Conversation. February 8, 2017.

100. Muncaster, Phil. "Russian Cybercrime Gangs Flourish with 1,000 New Employees" Infosecurity Magazine. November 23, 2015.

101. "German parliament cyber-attack still 'live'" *BBC News.* June 11, 2015.

102. Corera, Gordon. "Bulgaria warns of Russian attempts to divide Europe." *BBC News.* November 4, 2016.

103. Gayle, Damian. "CIA concludes Russia interfered to help Trump win election, say reports." *The Guardian.* December 10, 2016.

104. Kramer, Andrew. "How Russia Recruited Elite Hackers for Its Cyberwar." *The New York Times.* December 29, 2016.

105. *Ibid.*

106. Kramer, Andrew. "Online Attack Leads to Peek into Spam Den." *The New York Times.* September 2, 2013.

107. VK.com. https://vk.com/doc-94744292_437246143.

108. Knake, Rob. "Russian Hackers Were Only Getting Started in the 2016 Election." *Fortune.* January 15, 2017.

109. E-ISAC. "Analysis of the Cyber Attack on the Ukrainian Power Grid." E-ISAC. March 18, 2016.

110. Knake, Rob, *op cit.*

111. Nakashima, Ellen. "Russia has developed a cyberweapon that can disrupt power grids, according to new research." *The Washington Post.* June 12, 2017.

112. Karatzogianni, Athina. "Blame It on the Russians: Tracking the Portrayal of Russian Hackers during Cyber Conflict Incidents" *Digital Icons Magazine.* 2010.

113. Pagliery, Jose. "Ex-NSA director: China has hacked 'every major corporation' in U.S." CNN. March 16, 2015.

114. Sanger, David, David Barboza and Nicole Perlroth. "Chinese Army Unit Is Seen as Tied to Hacking Against U.S." *The New York Times*. February 18, 2013.

115. Denning, Dorothy. "Cybewar: How Chinese Hackers Became A Major Threat to the U.S." *Newsweek*. October 15 2017.

116. "Chinese hacker groups to shift focus to India in 2018: FireEye." *India Times*. December 7, 2017.

117. Sanger, David, David Barboza and Nicole Perlroth. "The World Once Laughed at North Korean Cyberpower. No More." *The New York Times*. October 15, 2017.

118. Sanger, David and Martin Fackler. "N.S.A. Breached North Korean Networks Before Sony Attack, Officials Say." *The New York Times*. January 18, 2015.

119. Kim, Sam. "How North Korea Built An Army of Hackers: Q&A." Bloomberg. October 17, 2017.

120. Shieber, Jonathan. "North Korean hackers stole South Korean and U.S. war plans." Tech Crunch. October 10, 2017.

121. Bosotti, Aurora. "North Korea bitcoin WARNING: Kim regime hacking cryptocurrency to fund nuclear weapons." Express. December 22, 2017.

122. Sanger, David and Martin Fackler. "N.S.A. Breached North Korean Networks Before Sony Attack, Officials Say." *The New York Times*. January 18, 2015.

123. http://www.spiegel.de/media/media-35679.pdf

124. Hesseldahl, Arik. "Here's What We Know About North Korea's Cyberwar Army." Recode. December 1, 2014.

125. Greenberg, Andy. "Hacking North Korea is Easy. Its Nukes? Not so Much." Wired. October 10, 2017.

126. Denning, Dorothy. "Iran's Cyber Warfare Program Is Now a Major Threat to the United States." Newsweek. December 12, 2017.

127. Harris, Shane. "Forget China: Iran's Hackers Are America's Newest Cyber Threat." *Foreign Policy.* February 18, 2014.

128. Denning, Dorothy. "Iran's Cyber Warfare Program Is Now a Major Threat to the United States." Newsweek. December 12, 2017.

129. Thompson, Mark. "Iranian Cyber Attack on New York Dam Shows Future of War." Time. March 24, 2016.

130. Jones, Sam. "Cyber warfare: Iran opens a new front." *Financial Times.* April 26, 2016.

131. Brunner, Jordan. "Iran Has Built an Army of Cyber-Proxies." The Tower. August 2015.

132. Sanger, David. "Obama Order Sped Up Wave of Cyberattacks Against Iran." *The New York Times.* June 1, 2012.

133. Santiago, Denis. "Cyber Warfare Iranian Style." Huffington Post. January 9, 2018.

134. Morgan, Steve. "Cybersecurity Market Reaches $75 Billion In 2015; Expected To Reach $170 Billion By 2020." Forbes. December 20, 2015.

135. Valette, Antoine. "The Threats of Hacking: Part 1." Kroll on Track. September 7, 2015.

136. Arthur, Charles. "How computer hackers do what they do … and why." The Guardian. June 6, 2011.

137. Curtis, Sophie. "Anonymous recruits amateurs in cyber war against Isil." The Telegraph. November 18, 2015.

138. Richmond, Riva. "How Anonymous Shut Down Sites." *The New York Times.* December 13, 2010

139. Peterson, Andrea. "Amateur hackers probably caused Friday's Internet meltdown, researchers say." *Los Angeles Times.* October 26, 2016.

140. Sherman, Ted. "How a Rutgers student went from Minecraft to internet warfare." NJ.com. December 17, 2017.

141. Zetter, Kim. "Hacking Team Leak Shows How Secretive Zero-Day Exploit Sales Work." Wired. July 24, 2015.

142. Goldstein, Matthew. "Need Some Espionage Done? Hackers Are for Hire Online." *The New York Times.* January 15, 2015.

Chapter 8

143. Fessler, Pam. "Timeline: Foreign Efforts To Hack State Election Systems And How Officials Responded Facebook Twitter Flipboard Email." NPR. July 31, 2017.

144. "2016 Presidential Campaign Hacking Fast Facts." CNN. February 5, 2018.

145. Fessler, Pam. "If Voting Machines Were Hacked, Would Anyone Know?" NPR. June 14, 2017.

146. Woodruff, Judy. "The bipartisan effort to protect future elections." PBS Newshour. January 11, 2018.

ABOUT THE AUTHORS

David Cahn and **Jack Cahn** are journalists, activists, and ethical hackers. The identical twins are authors of *When Millennials Rule*, which was featured on PBS, *Fortune, The New York Post* and *USA Today*. David and Jack have been featured in *The New York Times, Wall Street Journal, Time Magazine,* and *New York Magazine*. In 2017, the brothers graduated from the University of Pennsylvania with two Bachelor's Degrees each, in Computer Science and Economics.